# AIDS *for* AMATEURS

# AIDS *for* AMATEURS

*Human Choices,*
*Immune Responses, Social Burdens*


DONALD GENE PACE, PHD
AND
OMAR BAGASRA, MD, PHD

*AuthorHouse™*
*1663 Liberty Drive*
*Bloomington, IN 47403*
*www.authorhouse.com*
*Phone: 1-800-839-8640*



*Published by AuthorHouse 6/12/2013*

*ISBN: 978-1-4817-6124-6 (sc)*
*ISBN: 978-1-4817-6125-3 (e)*

*Library of Congress Control Number: 2013910141*



*This book is printed on acid-free paper.*

# Dedication

To our wives and children, whose patience and understanding have played a pivotal role in the development of our work.

OB/DGP

# Contents

# chapter one

# A Preventable Disease

SIDA (AIDS in Spanish)

| Bypass |
| --- |
| Threatening, menacing |
| Aiding, abetting, choosing |
| Individual decisions, global consequences |
| Avoid |

FOLLOWING THE SPANISH DISCOVERY of America by the Genoese Christopher Columbus, that nation, its Iberian neighbor Portugal, France across the Pyrenees Mountains, England nearby across the water, and other nations followed the great mariner's foot-in-the-door discovery with their own feet, legs, heads, and arms—those attached to the shoulders and those which European ammunition exploded on American soil. More potent and deadly than those menacing weapons, disease exploded on the New World scene. Although typically unintentional, various versions of "germ warfare" decimated native populations of the various Americas. This destruction-by-disease was an offshoot of the 12 October 1492 discovery, but it was not the fault of the Admiral of the Ocean Sea. The lack of understanding about how to prevent disease was as profound, and more deadly, as the profound geographical ignorance that early explorers and settlers ignorantly shared. Frustration accompanied fear as disease mysteriously massacred native peoples. Lack of disease prevention has plagued every society; only the levels of sickness and mortality vary. It is difficult to war against a visible enemy, but fighting invisible enemies is exasperating.

Fighting HIV has been both costly, and lethal. But unlike the futile fight against smallpox, and other diseases, that post-Columbian Native Americans routinely lost, the modern-day battle against AIDS is one that could have been halted early, and then stopped dead in its morbid tracks. HIV, the viral foot-in-the-door that leads to subsequent colonization and exploitation by AIDS, has been seen under microscopes by scientists, and has been slowed by antiretroviral cocktails (mixtures of drugs that fight the HIV retrovirus, and are known as HAART), but as of yet it has neither been prevented by vaccine nor cured by prescription.

The war against AIDS needs to be won, but it might best be thought of as a war that is won by avoiding the enemy. HIV is deadly once engaged. Why pitch to it, if it routinely hits home runs? Why let it into “the paint” if it scores with slam-dunk regularity? Engaging in behaviors that tempt HIV has proven no more logical than pitching fastballs to Barry Bonds, or leaving Shaquille O’Neal alone under the basket. Common sense still has its place. HIV is the mother of AIDS; human choice is the father of HIV. AIDS is, in reality, an illegitimate disease. If deprived of both father and mother, future illegitimate conceptions could simply be prevented.

## chapter two

# "LIKE A WORM IN THE VEGETABLES"

ghajnuniet (AIDS in Maltese)

| Perspective<br>Suggested, commanded<br>Thinking, feeling, praying<br>Science and religion partners<br>Vision |
|---|

AIDS IS, ULTIMATELY, A preventable disease but its prevention must come from correct adult human choices, not some magical scientific cancellation of consequences for bad behavior. The conduct that leads to HIV infection, and that indirectly afflicts innocent others who suffer through no bad choice of their own, is inconsistent with the major world religious traditions. In 1987, Ted Koppel, the famous and respected anchor of ABC Nightline, shared universal wisdom that has been much quoted since. It has become popular for the same reason that Tom Paine's *Common Sense* [1] appealed to American colonists at the time of the Revolution, or Harriet Beecher Stowe's *Uncle Tom's Cabin* [2] appealed to those who knew slavery was morally wrong and, through interaction with her book, solidified their own beliefs. Dr. Martin Luther King's famous "I have a dream" speech was all the more appealing because so many who did not share his eloquence did share his dream [3]. So it is with Koppel's statements at the 1987 Duke commencement: "We have actually convinced ourselves that slogans will save us. 'Shoot up if you must; but use a clean needle.' 'Enjoy sex whenever with whomever you wish; but wear a condom.'" He reminded the Duke audience that motives still matter: "No. The answer is no. Not no because it isn't cool or smart or because you might end up in jail or dying in an AIDS ward –but no, because it's wrong." He reminded his audience that "What Moses brought down from Mt. Sinai were not the Ten Suggestions, they are Commandments. Are, not were." [4]

Koppel's reminders are more than just Koppel's reminders; they echo teachings of major world religious traditions (including Islam, Judaism, Christianity, and Hinduism), which all revere the prophet Moses and the doctrines to which the news anchor referred. The eighth commandment recorded in the twentieth chapter of the Old Testament Book of Exodus [7] finds echo in the seventeenth chapter of the Holy Qur'an [5], which characterizes adultery as both evil and shameful. The Jewish Talmud declares that "Immorality in the house is like a worm in the vegetables" (*Talmud, Sota 3b*) [5].

AIDS has become that worm in many houses in many lands. Hinduism also decries adulterous behavior, both on moral grounds and on grounds of caste confusion. Vishnu Purana 3:11 warns against immorality with another man's wife, teaches that even immoral thoughts are evil, states that adultery brings both mortal and post-mortal punishment, and condemn the adulterous to later rebirth as an insect that creeps: "A man should not think incontinently of another's wife, much less address her to that end; for such a man will be reborn in a future life as a creeping insect. He who commits adultery is punished both here and hereafter; for his days in this world are cut short, and when dead he falls into hell" [5]. The Hindu Laws of Manu (Manusmriti) argues that "Men who commit adultery with the wives of others, the king shall cause to be marked by punishments which cause terror, and afterwards banish" [6].

Throughout history, positive behavior has been linked to adherence to community religious values. AIDS is not a disease that humans desire or plan to acquire; it is the result of irrational emotional behavior. Although the fear of suffering may deter HIV-inviting behaviors, compliance with the time-tested values of the world's great religious traditions provides an even more powerful deterrent. AIDS prevention efforts should not only focus on scientific logic but on the control of emotions. Because HIV prevention is a matter of both heart and mind, both religion and science should be utilized as partners in prevention rather than mutually-exclusive antagonists. Although AIDS commonly traces its roots to unholy origins, the crusade against AIDS is one in which the peoples of the world must unite. A divisive holy war will not suffice; the world needs a unifying jihad against a killer that threatens people of all faiths, political persuasions, racial characteristics, and geographic origins. It is a particularly tough war to win since human volition is involved, and since most infected men or women do not know they are infected, and over 80% of the underdeveloped nations do not have free HIV testing facilities. Nevertheless, when truly just and truly holy, wars can be worth fighting, and winning.

# chapter three

# TINY FEMALE VAMPIRES AND AIDS

VIGS (AIDS in Afrikaans)

| Myths |
|---|
| Classmates, mosquitos |
| Hand-shaking, pen-sharing, ball-playing |
| Ignorance bypasses crucial concerns |
| Distractions |

HIV INFECTION IS INTEGRALLY associated with the blood. Mosquitoes have taken a real interest in human blood. A mosquito bite occurs when a mosquito penetrates the human skin with what is called a proboscis, a type of miniature needle. Females, the only mosquitoes that bite humans (male mosquitos feel no biological need to do so), sense a potential victim by an elaborate sensing system that can detect odor, carbon dioxide, light, and heat. Blood typically coagulates quickly, and thus checks excessive human blood loss. Fortunately, from her viewpoint, mosquito saliva has anticoagulant proteins that keep human blood flowing by preventing clotting. Sucked blood enters the abdomen, and the undisturbed insect will continue ingesting the blood until her sensory nerve notifies her that the abdomen is full. Incidentally, a nonfunctioning sensory nerve would result in a blood-exploded mosquito. [8]

Could a female mosquito suck blood from an AIDS-infected

person, and then spread HIV-1, the AIDS virus, to its next blood provider? The short answer is "No." The longer answer, also "no," goes something like this: The amount of mosquito saliva is so small that it has no potential to infect another human. This does not mean that concern for saliva-borne infection has been unfounded. Such historically lethal megakillers as malaria, and yellow and dengue fevers, can be spread from person to person through mosquito saliva. Although humans have far more saliva than mosquitoes, and although HIV has been documented in saliva, it has only been detected in exceptionally low amounts. Human saliva, by itself, has never been the means of HIV transmission. As disgustingly mean as spitting may be, it is not a means of transmitting HIV from spitter to spite, nor is it spread through kissing, when a larger saliva exchange occurs between mouths. [9]

In spite of extensive research, the key institutional complex known as the Centers for Disease Control (CDC) has reported zero documented cases of AIDS transmission via a mosquito, or any other insect, for that matter [167, 17]. This applies to areas that have plentiful supplies of both HIV-infected humans and mosquitoes. These biters, and other insects, do not share human blood or their own blood with subsequent bitees. When a female mosquito "bites" her victim, she does not share blood but rather injects a tiny quantity of lubricating saliva that enables her to feed more easily. HIV survives only briefly inside mosquitoes, and cannot reproduce. There are no HIV-positive mosquitoes or insects. As deadly as they may be in passing along numerous other pathogens, they have no capacity to infect humans with HIV.

What about blood that remains on a mosquito's mouth parts; can it cause HIV infection? This is nothing to worry about. Mosquitoes rest and digest before looking for a new involuntary blood donor. By the time they extract more blood, they have no living HIV to share. Moreover, tiny mouth parts can host only tiny amounts of blood, and this, too, works against HIV-sharing by mosquitos [9].

Recent times have witnessed new threats to life, which have resulted in death by automobile accident, death by drug overdose, premature death by overnourishment in a world plagued by malnutrition. There have been numerous deaths influenced by cars, bombs, and guns; and numerous premature deaths indirectly induced, indirectly cheered on, by overnourishment, by excessive sugar intake, by mindless addiction to frivolous video games, and by lethargically immobile attendance (in front of ever larger video screens) at games that feature the vigorous physical exertions of talented, but at times steroid-loaded, athletes. Yet it is critical to remember that cars, bombs, guns, food, sugar, and televised sports are not the ultimate cause of human suffering and death; humans are. It is they who drive the cars, drop the bombs, pull the triggers, feed themselves beyond reason, promote sugar cravings, and volunteer to be couch potatoes. Still, historically most deaths likely have not come through human choice, but through human incapacity to counter disease. The mosquito has been a leading actor on this stage of human life, and has unwittingly removed millions from the stage: death by malaria, death by yellow fever, death by diarrhea, death of even very large persons by tiny mosquito proboscises.

The mosquito's needle-like "beak," its blood-drawing instrument, intentionally extracts blood, and unintentionally infects with malaria and yellow fever, but not HIV. This tiny nemesis to human health seeks human blood for personal nourishment, yet has caused and continues to cause, untold human suffering. Yellow-fever, malaria, dengue, and West Nile virus are among the deadly examples of the power these miniature vampires have exerted over human life and history. Their effects are brutal yet their motives appear quite innocent: female mosquitoes extract blood to help them produce eggs (males have no such need, and do not drink blood). The change of the exterior environment can menace mosquitos, can stop their proliferation, and even stop their lives. For those who do become sick with malaria, it can be fatal but not always. It will, whether fatal or not, be miserable. HIV is not spread through the external environment, including through mosquitos that share that environment. The largely private world in which HIV spreads is not the world of mosquitos.

HIV/AIDS is generally the result of human choices, although there are numerous innocent victims who had no choice in their own fate. For those who take the risks, who break the rules, it is not merely a matter of fate, but of selfishness, of unfaithfulness, of risk-taking, of exploitation, of disregard for societal norms, political laws, and moral values. It is behavioral fascism. Mosquitoes are active on other fronts, but they are not a direct threat on the AIDS killing fields. Humans are.

Mosquitos, blood, and malaria; mosquitos, blood and yellow fever; mosquitos, blood and dengue. Why not mosquitos, blood, and human immunodeficiency virus? Mosquitos can perform their blood-extraction routine just as easily with humans who are free from HIV as with those who are not. Mosquitos do not discriminate in such matters. But whether a mosquito robs the blood of a totally uninfected person, one with a dormant HIV infection, or one with a full-blown case of AIDS, the HIV-transmission result is the same: *nada*. The mosquito cannot transmit HIV to humans. The quantity of infected blood ingested by the insect is simply too limited to allow for human-mosquito-human transfer of the HIV retrovirus. The very size of the mosquito, which facilitates its stealthy capacity to exact involuntary tribute-in-blood from unsuspecting victims, proves a decisive disadvantage in spreading AIDS. Troublemaking-via-mosquito, fortunately for humans, has its limits. Mosquitos are, as it were, kept behind bars when it comes to spreading HIV.

In San José, Costa Rica, a tour guide commented on the proliferation of protective metal bars in front of house after house, and business after business: "We have a big problem with mosquitos." Participants in the tour, including one of the authors of this book, laughed, and then heard the tour guide speak of two-legged mosquitos that the metal bars are designed to deter. As useful as these unaesthetic bars may be in curtailing crime, more (not fewer) bars are needed to fight unwelcome intruder known in Spanish as SIDA (síndrome de inmunodeficiencia adquirida; AIDS). Bars of societal norms, bars of political enactments, and bars of moral values must be reinforced, not weakened, by self-defeating approaches, if the two-legged mosquitos that cause AIDS are to have their wings clipped.

# chapter four

## "As a Whale goes Through a Net": Getting to the Roots

زدي إلا (AIDS in Arabic)

| Leviathan<br>Imposing, mighty<br>Menacing, attacking, breaking<br>Stronger net, weaker whale<br>AIDS |
|---|

PUBLIC POLICY CONCERNS REQUIRE public leadership, and a narrow focus on branches rather than roots will not get to the root of public dilemmas. It may not deal effectively with the branches either. In a notable editorial in the *Wall Street Journal* titled "The Joy of What?" [10], editorial authors argued in favor of greater public openness in discussing root causes of serious social problems: "Sin isn't something that many people, including most churches, have spent much time talking about or worrying about through the years of the revolution [in moral behavior]. But we will say this for sin: it at least offered a frame of reference for personal behavior. When the frame was dismantled, guilt wasn't the only thing that fell away; we also lost the guidewire of personal responsibility." The very process of debating an issue can have a deterrent effect by persuading humans to accept the human responsibility for moral agency. Human can choose, they can think, but they must think in a multidimensional fashion. HIV/AIDS, like other complex problems, is, by nature, complex. It is not simply a biological issue or a poverty issue or an economic issue. It is all of them, and much more, including a moral issue that is often related to right and wrong choices. The *WSJ* editorial states: "The United States has a drug problem and a high-school-sex problem and a welfare problem and an AIDS problem...None of this will go away until more people in positions of responsibility are willing to come forward and explain, in frankly moral terms, that some of the things that people do nowadays are wrong."

Regardless of how AIDS has been publically packaged, for example as an unfortunate biological phenomenon or as an expensive burden to taxpayers, it does have religious and moral dimensions. D. Todd Christofferson, now a notable clergyman but earlier a young law clerk to Judge John G. Sirica [11], who led the "Watergate" prosecution against President Richard Nixon, has expressed concern over the lack of moral guidance given to young people. Youth should not, he argues, be left to themselves "to understand and evaluate the alternatives that come before them." Without moral direction about right and wrong, they may be inadequately prepared to face the "vigorous, multimedia advocates of sin and selfishness." These negative forces do not simply leave youth alone to make intelligent decisions; they seek their own, often avaricious, agendas. In 2009, Elder Christofferson lamented, "The societies in which many of us live have for more than a generation failed to foster moral discipline. They have taught that truth is relative and that everyone decides for himself or herself what is right. Concepts such as sin and wrong have been condemned as "value judgments" [12].

The apostle Paul, whose writings comprise much of the Christian New Testament, wrote that "the wages of sin is death" (Romans 6:23) [7]. While his statement has strong spiritual connotations (death involves separation; sin separates one from God), it may be drawn upon for wording that expresses the link between private moral decisions and suffering from biological consequences. To those who ignore common sense and common religious teachings in such common religious traditions as Islam, Judaism, and Christianity, HIV/AIDS may become a deadly problem. Societies are incapable of controlling moral, or immoral, behavior without moral teachings; indeed, many seem to have abdicated this duty and privilege. Sin, a central concept in Islam, Judaism, and Christianity, flourishes in a moral vacuum; so does HIV. Although HIV infection often occurs among guiltless individuals (including innocent newborns, and blameless spouses), it is, in its roots, related to moral behavior that is characterized as sin by the world's major religious traditions, and most minor traditions as well. Public policy makers, parents, teachers, and others that influence those public moral values that guide private moral decisions, have the opportunity to influence behavioral roots, and not simply complain about menacing branches.

Rhetoric, the art of persuasion, is an art that needs to be turned toward roots rather than branches. Speaking about what has become the most influential constitution in the world, US President John Adams explained: "Our constitution was made only for a moral and religious people. It is wholly inadequate to the government of any other." As proud as he was of the new nation he helped bring into being, Adams admitted that his country had "no government armed with power capable of contending with human passions unbridled by morality and religion. Avarice, ambition, revenge, or gallantry, would break the strongest cords of our Constitution as a whale goes through a net" [13]. Because of the contemporary confluence of poor human decisions and the genetic mutation of retroviruses, AIDS has become a whale that breaks one after another net of resistance. The net can be strengthened and the whale weakened, but not without moral values, the right to proclaim those values, and the courage to do so, the courage to tackle deep-rooted problems with moral persuasion, the courage to use rhetorical skills to protect the body politic.

# chapter five

# A Tale of Incomplete Hopes

СПИД (AIDS in Russian)

| Methodology |
|---|
| Evasive, dormant |
| Invading, mutating, hiding |
| Complicated attacks, complex solutions |
| Paradigm |

IN A SUPPOSEDLY PATH breaking August 2011 article in *Science* Express, Xueling Wu, et al. [14] described meticulously the process by which HIV neutralizing antibodies develop. Scientists from the NIH's Vaccine Research Center (VRC) led the investigative effort that provided what the NIH boldly announced were "vital clues to guide the design of a preventive HIV vaccine" [15]. At the heart of the research were three antibodies (Abs): VRC01, VRC03, and VRC-PG04. All three Abs were found in the blood; the first two came from a North American donor, and the third from an African donor. Both donors were HIV-1 seropositive. The HIV-binding segments of the three Abs have shown remarkable potential for neutralizing the HIV-1 retrovirus. Although the three antibodies have structural similarities as well as differences, their similar regions permit the three to contribute collectively to HIV-1 neutralization by binding "to the same spot on the virus." By so doing, the three Abs are able "to neutralize a high percentage of HIV strains from around the world" [15].

According to NIAID (National Institute of Allergy and Infectious Diseases) Director Anthony S. Fauci, "This elegant research brings us another step closer to an HIV vaccine and establishes a potent new technique for evaluating the human immune response to experimental vaccines, not only for HIV, but for pathogens generally" [15]. This most recent article pursues the same path of a 2010 discover by VRC researchers that proclaimed the HIV-neutralizing power of the Abs they were showcasing. In fact, they announced, two of the antibodies "could stop more than 90 percent of known global HIV strains from infecting human cells in the laboratory" [15].

The North American donor (donor 45) provided antibodies that were labeled VRC01, VRC02, and VRC03, obviously named after the Viral Research Center (VRC) team who discovered and reported them publically. This was impressive, but even more impressive was the 2011 VRC finding that they had discovered antibodies analogous to VRC01 in the blood of donors 74 and 0219, both from Africa.

Because of the momentous discovery "that these VRC01-like antibodies all bind to the same spot on HIV in the same way," it would seem imperative that any future attempts at HIV-1 vaccine design focus heavily on including, in the vaccine "a protein replica of this spot, known as the CD4 binding site, to elicit antibodies as powerful as VRC01." The constantly mutating HIV retrovirus has baffled researchers who have had considerable difficulty dealing with the perennial mutations. In contrast to the general mutational patterns of HIV, CD4 provides something predictable, a binding site that remains constant among the many HIV varieties across planet. When seeking to infect a cell, the retrovirus consistently utilizes this CD4 binding site.

The VRC01-like antibodies have genes that are highly mutagenic: between 70 and 90 mutations "between the first draft that codes for a weak antibody and the final version that codes for an antibody that can neutralize HIV." Because the genes are found in the DNA of B lymphocytes, a type of immune cell,

VRC director Gary J. Nabel, explains that vaccine research must focus on these B cells: "To make a vaccine that elicits VRC01-like antibodies, we will need to coach B cells to evolve their antibody genes along one of several pathways, which we have now identified, from infancy to a mature, HIV-fighting form" [15].

Through creative use of a methodology known as "deep sequencing," the Wu group tracked "the evolution of the antibody response to HIV at the genetic level." As Peter Kwong, a co-principal investigator of the 2011 article optimistic noted: "We found a way to read the books, or genes," in the large genetic library they had pulled together through deep sequencing, and following sophisticated analysis, were able to define the distinguishing characteristics that VRC01-like antibodies exhibit. Co-PI John R. Mascola indicated that it will now be more feasible to assess, at the preclinical or clinical levels, whether or not a proposed HIV vaccine is moving in the correct direction. "As we develop and test new HIV vaccines," he noted, "it will be possible to analyze not just antibodies in the blood, but also the specific B-cell genes that are responsible for producing antibodies against HIV" [15].

This type of announcement, while it sounds hopeful, tells only part of the story of fighting HIV/AIDS. Stimulating antibody reaction is not enough. HIV remains dormant; it hides within cells, and is not subject to the same antibody defenses that clear up other diseases. Over 40 species of non-human primates are infected with SIV, similar to HIV in humans, and yet their infection does not progress to AIDS or other life-threatening conditions associated with immunodeficiency. We maintain that, as exciting and profitable as traditional antibodies-will-someday-defeat-AIDS research may be, traditional innate and acquired immunities simply have not succeeded in stopping AIDS. We further maintain that there is another means of defense that does stop HIV from progressing to AIDS. It is microRNA, something that can stop HIV within the cell, at a molecular level.

# chapter six

# To Fund or Not to Fund? That is Not the Whole Question

СНІД (AIDS in Ukrainian)

| Money |
|---|
| Helpful, partial |
| Preventing, treating, equalizing |
| Amoral teaching, poor returns |
| Inadequate |

GOVERNMENT FUNDING CAN PLAY a vital role in each of the three areas emphasized by President Barack Obama: prevention, treatment, and reduction of health disparities. Standing in the East Room of the White House on 14 July 2010, President Obama outlined his administration's three-pronged approach, and made a plea for greater attention to groups that suffer disproportionately from HIV/AIDS: "We all know the statistics. Gay and bisexual men make up a small percentage of the population, but over 50 percent of new infections. For African Americans, it's 13 percent of the population -- nearly 50 percent of the people living with HIV/AIDS. HIV infection rates among black women are almost 20 times what they are for white women. So, such health disparities call on us to make a greater effort as a nation to offer testing and treatment to the people who need it the most." [16]. The President reminded his listeners that he realized "this strategy comes at a difficult time for Americans living with HIV/AIDS, because we've got cash-strapped states who are being forced to cut back on essentials, including assistance for AIDS drugs. I know the need is great. And that's why we've increased federal assistance each year that I've been in office, providing an emergency supplement this year to help people get the drugs they need, even as we pursue a national strategy that focuses on three central goals" [16].

More money certainly can help, but money is not the whole answer, nor is it the best answer. The teaching of private moral values in homes, churches, schools, and other settings need not be unduly expensive; indeed, it may be free, or virtually free. A return to the basic values of moral behavior taught by the leading religious groups in the US, and world, would be a cost-efficient means of tackling a public policy dilemma that even generous government support cannot tackle successfully. During this time of "cash-strapped states who are being forced to cut back on essentials, including assistance for AIDS drugs" [16], it is crucial that governments should back measures that support institutions, especially the traditional family, that can assist in fighting HIV/AIDS. Public and governmental support for measures, such as the Defense of Marriage Act, is a step in the right direction. It is important to acknowledge reported figures, examine trends, and probe for answers. Logical solutions, however, are much easier in the abstract than in real world settings. Broken homes and broken communities surely play a part, as do discrepancies in educational opportunities, and lack of religious instruction. Poverty, health care disparities, and entrenched racial biases all contribute as well. Yet these seemingly intractable challenges may be at least as hard to change as the behaviors that cause HIV infections. The means to prevent HIV are largely known, yet because they intersect with the human freedom to choose, even the best efforts to prevent new cases have often proven as elusive as the quest for an AIDS vaccine.

## chapter seven

# The Human Immune System: Multi-Level Protection

UKIMWI (AIDS in Swahili)

| Immunity<br>Antibodies, lymphocytes<br>Sensing, responding, failing<br>Classical immunity proves inadequate<br>microRNA |
|---|

A KEY WORD ON which the acronym AIDS is based is "immune," and a basic understanding of the immune system is essential in order to appreciate the challenge HIV presents to that remarkable system. The human immune system prevents the entry of tiny but dangerous entities that can make us sick (pathogens), and fights against them to eliminate the threat when such an invader does get inside the body. As simple and commonplace as it seems, the skin is of enormous value to humans. Dangerous viruses and bacteria, which could make us sick, commonly are stopped by this first line of defense. If these threats do enter the body, antibodies can form to counter them, and the body has a memory system to remember such invaders on later occasions. Even at the level of tiny molecules, protection takes place. The different components of the immune system work in harmony with each other. Much is still not known about how these components work, but what we do know causes us to marvel at the complexity, and yet overall simplicity, of the multi-level protective system we call the immune system. Some further details are both informative and interesting.

What we inherit at birth is called innate. Basically, it stays as it is, and it is not changed due to its interaction with different pathogens. This is far different from what is called the classical (or adaptive) immune system. The two systems, innate and adaptive, function well together in a concerted effort to fight against pathogens. The innate system, including the skin, initially keeps infection from harming the body, or at least delaying its damage. This is a type of "buying time," so that the adaptive immune system can carry out its defensive work. Features of the body that give basic protection are the skin, which serves as a barricade to most attacks by pathogens, and other responses, including reflexes such as sneezing or coughing or getting rid of pathogens through secretions that leave the body. Tears flow from the eyes, and help kill what we call "germs." Beneficial microbes called normal flora help provide protection also. These include fungi and bacteria, which remain in body cavities such as the mouth, the female vagina, and our 36-feet-long gastrointestinal pathway all the way down to the anus. There is a biofilm in these areas that creates a protective environment. We generally take these features of the immune system for granted, unless we have a particular problem that shakes us from our complacency.

Innate immunity dedicates itself to rapid response; it works quickly, even within seconds to counter an infection. Adaptive immunity, on the other hand, can take days or even month to successfully meet a challenge. Although much slower than innate immunity, adaptive immunity is remarkable for its capacity to "remember" past attacks. This archiving, and recalling, of the body's own history is comparable to an experienced general's ability to profit from previous experiences. Like an experienced general, adaptive immunity gets better as time passes. Adaptive immunity deals with far more specific threats than innate immunity. Still, neither is inherently better than the other; each is remarkable, and as a team they complement and support each other.

Part of the fascinating work that the immune system performs is done by useful cells known as scavenger cells (including phagocytic and endocytic cells). These scavengers learn to sense dangerous molecules, and then overwhelm them. They learn what is not only "foreign," but also "dangerous." Just as not all immigrants are criminals, but some are, the scavenger cells learn to sense both foreign and danger. Food, for example, would typically not be regarded as dangerous, even though it is foreign, similar to migrant workers who help us harvest the crops. The scavenger cells help tidy the body up, to keep it free of pathogens and other unwanted clutter [125].

Another key component of the adaptive immune system is called complement. This system provides the initial, and at times even immediate, response to pathogens that can multiply and spread through the body's fluids. Complement molecules are present in blood plasma.

The complement system affords the initial, even instantaneous, anti-pathogenic innate response to counter those pathogens capable of proliferating in bodily fluid. Complement molecules are found in blood plasma. They fasten themselves onto bacteria, and they can break through the wall of a bacterium's cell wall and membrane. This breaks, or ruptures, the basic structure of the bacterium, which leads to its destruction. It is similar to poking a knife through a ripe watermelon. This is termed lysis. Another approach that complement takes is to coat a particular bacterium, which prepares it to be destroyed by other protective elements called neutrophils and macrophages.

The word *phago* has reference to the eating process; the word *cyto* has reference to the cells. The term *phagocytic*, therefore, signifies a process in which cells that are dangerous or not useful are eaten-up by monocytes [125]. Phagocytic cells eat fungi, bacteria, and parasites. A type of macrophages called monocytes flow within the blood stream, migrate to the different organs in the body, and play the role of a hungry garbage collector that eats non-functional cells and those that have died.

These phagocytic cells eat bacteria, fungi, and parasites. Therefore, cells called monocytes circulate in the blood stream, and are supposedly non-differentiated macrophages. These cells migrate to various organs, and carry out a garbage-cleanup function by eating dead cells, and non-functional cells (phagocytizing), that have aged or damaged for some reason.

Another type of phagocytic cell type is the neutrophil. These far outnumber the macrophages, and they multiply exceptionally quickly. In fact, they multiply faster than any other cell type in the body. They have a very short lifespan, only between about two and four hours. In the peripheral blood cells, they are the most common type of white blood cell. Their presence in bone marrow is crucial to human health, and even human life. At each immune response phase, neutrophils are critically important. They have receptors that attract bacteria that complement has already worked with. The neutrophils keep us alive; without them, attacks by bacteria would be deadly.

An especially important time to have neutrophils functioning well is when a person undergoes chemotherapy, which is commonly used to check the growth of cancer cells. The genius of chemotherapy is its ability to identify, and destroy, proliferating cells. This is why hair falls out, and fingernails are affected; they are cells that have been identified, and then damaged or destroyed. Life is at stake if the neutrophils are threatened, as they are during chemotherapy, but thanks to modern medical practices, production in the bone marrow of neutrophils (and also the helpful macrophages) can be stimulated by medication. Chemotherapy can destroy bad (cancerous) cells, while damage to good cells can be minimized. After chemotherapy or radiation is completed, an injection of GMCSF is given (granulocytes/ macrophage stimulating factor; marketed and sold as *Neupogen, Filgastim, Sargramostim, and Neulasta*). This is a critically-important injection that preserves both neutrophils and life itself. It is also an often overlooked component of chemotherapy treatment.

To enter the body, HIV appears to bypass the innate immune system. Adaptive immunity tries to attack and defeat the virus, but the typical antibody response fails to succeed with HIV. Like a Trojan horse, HIV uses the immune system to destroy itself! Classical immunity, which is routinely helpful against pathogens, does not work against HIV.

Two categories of immunity, Antibody-mediated (humoral immunity) and cell-mediated (CMI), and the different vaccines that rely on these types of immunity, simply have not worked. Classical immunity relies on its ability to detect "any foreign" substance, and its capacity to thwart its intent to damage the host body [168]. This can be deadly when HIV is involved because HIV utilizes the immune system to activate, proliferate, and eventually bring death. Within the cells, a type of immunity known as "molecular immunity" utilizes exceptionally tiny strands of RNA known as microRNAs (miRNAs) to disable viral threats [127-129, 131, 158-161]. Vaccine attempts have routinely focused on classical immunity, but since these have not succeeded, it would make sense to try focusing on molecular immunity in vaccine research (our articles).

Classical immunity depends heavily on its use of lymphocytes. This system, the lymphocyte recognition system, senses that an invading substance is "foreign," and then prompts the creation of antibodies (Abs) or cell-mediated immunity [162-163]. When a virus or a special type of virus called a retrovirus (such as HIV), a bacteria, or some other "foreign" entity enters a person (or host), classical immunity makes its protective response. However, classical immunity, generally speaking, it limited to recognizing invaders that are outside of individual cells, or extracellular [162-164]. Although health preserving and useful in fighting extracellular pathogens, unfortunately classical immunity is not adept at opposing intracellular pathogens, or pathogens that have invaded the cellular genomes. Once a retroelement, such as HIV, is inside a cell, classical immunity is typically powerless to block its impacts. Retroviruses and other genetic parasites are able to enter and leave the human genome (the composite of all genes in the body) without even being detected through the mechanisms of classical immunity, but not so with miRNA, which provide immune protection within the cells themselves at intracellular levels [165]. In recent years, discoveries regarding miRNA have excited the scientific community, and raised hopes that they may hold a key to fighting HIV and other infections. Neither antibody-mediated immunity nor cell-mediated immunity has been able to deal effectively with *mycobacterium tuberculosis*, which is an organism that is covered by a specialized sheath that helps shield this potentially deadly organism from the protection generally afforded by classical immunity.

As tuberculosis has mutated, and drug-resistant varieties have developed, classical immunity has had even more difficulty countering this age-old menace that primarily attacks the lungs. Malaria, another long-time threat to human health, spreads through parasites that proliferate intracellularly (inside the cells) in the red blood cells of the host. The unfortunate result is the annual death of over one million children, not to mention the 1.2 billion people that are sickened by *malaria*. [132]. HIV infect target cells, including the CD4+ cells that are important in classical immunity, but these dangerous viruses persist in a dormant state until they have the opportunity to divide [128-131]. It seems terribly ironic that the very CD4+ cells that customarily protect against disease are "tricked" by HIV retroviruses into helping them multiply, and eventually even cause the death of the CD4+ cells themselves. With vastly reduced numbers of these protective cells, the body can no longer fight off other illnesses, such as pneumonia, with its usual effectiveness. In this way, the retrovirus HIV brings about the premature death of humans. We maintain that classical immunity will not be able to defeat HIV since it has been unable to conquer malaria, TB, listeria, leprosy, and other intracellular-induced illnesses. Not only does classical immunity lack the ability to reliably recognize invasions by HIV viruses, but it can also hinder healing by producing antibodies that end up assisting the proliferation of deadly retroviruses [127-128]. In short, classical immunity is simply not powerful to act within infected CD 4+ cells [126-130, 135, 157], and it actually plays a role in spreading HIV's destructive influence [136].

Cell-mediated immunity counters a virus by dispatching two types of cells that mature in the thymus (or T cells): CD8+ and CD4+). When these two types of T cells travel to places of trouble, cells that are producing viruses, for instance, the T cells can actually help to produce HIV. Fighting thus becomes assisting. It is as if friends become enemies, as T cells are transformed from trusty allies to traitorous foes. CD8+ T cells wipe out CD4+ T cells, and previously-uninfected CD4+ T cells promote a deadly cycle of destruction, bringing death first to immune cells and then to the infected persons themselves [137-138]. This explains the vital role that miRNAs must play. What an incomplete picture of the immune system we see if miRNAs are not included. These remarkable entities can actually block HIV growth within CD4+ T cells and also in macrophages [128, 131, 133, 139-141, 160, 165]. Put simply, miRNAs afford protection that adaptive immunity cannot give, including in the fight against AIDS. Recently, the astonishing potential of miRNAs has been explored, including in the battle against cancer, and against myriad viruses and microorganisms. We anticipate that they will be used routinely in a variety of therapeutic regimens.

Is it time to look beyond classical immunity? In the years following the onset of the AIDS crisis, classical immunity appeared logical as the principal foundation for vaccine research, but in the later years of the 1990s, theories of immunity emerged that focused on small dsRNA [127, 129]. Creatively drawing on observations of worms and plants, this type of immunological theory about RNA interference (RNAi) looked at how RNAi can inhibit viral proliferation in plant viruses through what is known as gene silencing. Moreover, there are applications throughout eukaryotic life forms [128, 131, 160]. As a result of their trailblazing work in RNAi, two US scientists (Andrew Fire and Craig Mello) received a 2006 Noble Prize [166].

Logically, researchers used traditional tools to analyze the emerging AIDS epidemic, but they underestimated the severity of the challenge. Traditional investigative tools were found to be seriously inadequate [127-128, 133-141, 156]. Basing their findings on investigations of humans infected with HIV-1, as well as macaques infected with SIV, researchers trusted in the ability of classical immunity to somehow win the immunodeficiency battle [133, 135-141]. Disappointingly, research projects were not able to show that traditional immunological therories held the answers to stopping HIV progression, or curing AIDS once it had begun [127-130, 135-137, 142, 143-145]. Vaccines have not materialized, and many trials, sometimes accompanied by the high hopes that media exposure brings, proved disappointing [127, 129, 135-140, 146-155]. Cumulative research is valuable, sometimes for what it shows to be true, and at other times for what it proves does not work. The vast research to date gives evidence that a paradigm shift is needed in HIV/AIDS research.

# chapter eight

# Strain 1, Group M, and the Epidemiology of HIV

СПИН (AIDS in Bulgarian)

> Instability
> Strains, groups
> Infecting, proliferating, baffling
> Subgroups strain global populations
> Mutations

HIV IS AN ORGANISM that evolves with exceptional rapidity. At the surface level, HIV may seem to be a single virus, but such is not even close to being the case. HIV not only has various subtypes that are genetically distinct from each other, but it is also remarkably subject to recombination. HIV-1 evolution in particular, has plagued humanity, including so-called "within-patient" viral adaptation. The plural term HIVs, not simply HIV, is the reality that the overall term HIV masks. Within just a few decades HIV/AIDS has gone from being unknown to assuming its place among the very most significant infectious diseases in the world, with a total number of cumulative infections now estimated at about 60 million. As of the end of the year 2009, the number of individuals living with HIV was an estimated 33.3 million, of which 2.1 million were children. Although HIV new infections seem to have reached their summit in 1996, the total number of persons living with the infection has continued to rise, in part because of the success of antiretroviral drugs. Uneven rates of prevalence, and varied epidemiological patterns characterize these infections. Resistance and adaptation to antiretroviral drugs further complicate matters [178]. Dealing with HIV is like walking on shifting sands.

Researchers who study epidemics, epidemiologists, have identified major HIV strains, groups, and subgroups. The predominant viruses are HIV-1 and HIV-2. The latter has two groups associated with it, A and B, and is endemic in the western part of the African continent. The global epidemic, so serious that it is often referred to as a pandemic, is almost entirely the result of the spread of group M (or Major Group) of the HIV-1 virus. In epidemiological term, this is the subtype or clade and group that have been most "successful." Epidemiologists subdivide this particularly virulent group M into nine subtypes, all of the consonants between A and K (A, B, C, D, F, G, H, J, K). As if this were not complicated enough, over 40 CRFs (circulating recombinant forms), which tend to have evolved more recently, add to the shifting sands of HIV complexity. The HIV/AIDS epidemic has been traced primarily to four subtypes (A, B, C, D) and two CRFs (CRF01_AE and CRF02_AG) [178].

Newly infected persons have the greatest capacity to infect others. Unfortunately, this comes at a time when "infectors" are unaware of their infection. The reason is that in the early phase of infection (about day 1 to day 10) the virus level in the blood is very high. After this generally short phase, the viral level comes down drastically. This happens apparently in the absence of any classical immunity. At high viral levels in the blood and semen, the infected person is very infectious and can spread HIV to other intimate partners and, if pregnant, to her unborn baby. At this phase, even if predisposed to do so, they are not likely to see a need to warn potential "infectees." Besides, persons who spread HIV to others often do so because they act according to emotion rather than using rational common sense. The main means that have helped to fuel the current HIV epidemic are sexual activity, injection with contaminated needles, , and transmission from mothers to their children while these mothers are pregnant, during the delivery of their babies. , Breastfeeding is generally not considered a risk for the newborn. Transmission via sexual contact accounts for roughly 80% of the HIV infections worldwide, and most of these contacts are heterosexual. An additional 10% of the world's HIV infections are due to injection drug use, or IDU. Infected individuals who are responding well to antiretroviral treatments have a substantially lower probability of infecting someone else than if they were receiving no such treatments (this is simply due to very low levels of virus in their blood). Likewise, male circumcision reduces the chances of a male becoming infected, which indirectly lowers female infections. The use of sanitized or clean needles, like the correct use of condoms, bring the likelihood of transmission to nearly zero [178].

What is HAART and what impact has it had? HAART stands for highly active antiretroviral therapy. Since its introduction in 1996, it has had a profound effect in lengthening the lives of HIV-infected individuals who previously were doomed to a rapid demise due to the progression of AIDS, and then death due to an AIDS-related cause. HAART uses a combination, or cocktail, that includes three different drugs that come from a minimum of two different classes. HAART has been so successful that AIDS has often become a chronic infection that is treatable, and that does not lead rapidly to death. Besides cost and distribution problems, which are considerable, HAART and its users have contributed to drug resistance problems. HIV, which mutates easily even without encouragement from patients who fail to adhere to prescribed routines, has undergone numerous mutations, and a serious problem known as TDR (transmitted drug resistance) has emerged [178].

The AIDS epidemic began in Africa, the Americas and Europe before it commenced in Asia, but given Asia's huge population, and the predictions of AIDS proliferation experts predict that the world of the near future will have three Asian nations with the greatest total number of AIDS cases: China, India, and Indonesia. In China, the world's most populous country, injected drug use was initially the leading cause of HIV infection. Later the main path of infection became, and still is, sexual contact. The epidemic in this country was primarily centered in high-risk groups, but evidence now indicates that the general population has become a serious part of the problem. HIV infection rates for drug users exceed 50% in some of China's worst areas for HIV/AIDS. In these highly troubled areas, women are also experiencing high prevalence rates [181].

In terms of the HIV/AIDS epidemic, the pattern seen in China is all too familiar. High-risk behavior in concentrated pockets of individuals initially accounts for most cases. HIV infection then becomes more generalized, and heterosexual activity in the general population comes to be a major part of the problem. China's self-imposed troubles with gender selection, and the resultant lack of females, will likely intensify the populous nation's challenges relating to AIDS. China and India, each with over one billion people, need to control HIV infections, for their own sake and that of the entire planet. What used to be seen, with some justification, as mainly an African problem, with some serious troubles in Europe and the Americas, is no longer a justifiable position to take. It now seems reasonable to assert, and worry, that as China goes, so goes the world; and as India goes, so goes the world. These two countries are home to about one-third of the world's inhabitants. HIV infection has truly become a global problem.

# chapter nine

# As Simple as ABC or ABCC?

АИДС (AIDS in Serbian)

Prevention
Traditional, widespread
A, B, C
Add one more C
Security

COUNTRIES AND ORGANIZATIONS WORLDWIDE have stressed three easy-to-remember basic prevention practices, the ABCs of controlling the deadly HIV retrovirus: Abstinence, Be Faithful, and Condoms [28]. While the ABC approach seems logical and simple, and may have been helpful in saving many lives, it has not proven to be either universally logical to researchers or appropriately simple to learners. In 2006, the Washington Post reported that the approach known as ABC, an "AIDS-prevention strategy widely promulgated both here and abroad, got a distinctly mixed report card as African countries reported their experiences to delegates at the 16th International AIDS Conference [in Toronto]" [32]. A study of black youth of junior high school age found that these youth were actually less likely to instigate sexual relations if taught only pre-marital abstinence than if given a more full A-through-C curriculum on AIDS prevention. A study of safe anti-HIV practices in Botswana found a disturbing disconnect between knowledge and practice. CDC-sponsored efforts to educate seem to have succeeded in raising educational awareness but not in changing actual behavior. The director of the joint US-Botswana collaboration, dubbed "BOTUSA," reported that "People who were exposed to the program had greater knowledge but were no more likely to be practicing ABCs." The ABC approach, on the other hand, seems to have been much more successful in Uganda [32].

In contrast to a mere hand-wringing recitation of statistical HIV woes, we suggest that the helpful ABC guidelines should be carefully adapted to the specific groups being served. Perhaps a group of middle school Hispanics needs less be-faithful-to-your-spouse encouragement, and a decreased emphasis on condom use, than instruction on the moral and practical values of postponing sexual relations until marriage. While many will already have had such intimate experiences, and even more will likely have them as they become older teens, they may well have fewer such experiences, and be more careful in partner selection than if they received no instruction. Some will even practice full abstinence. The dilemma of under-versus-over instruction must be faced wisely by local teachers and administrators. Too little instruction could prove deadly while excessively explicit instruction may actually inflame already highly inflammable emotions.

ABC, then, should be used judiciously. We propose that a second C, for Circumcision, should also be considered by educators, health care providers, and policy-makers alike. The titles of some of the now substantial literature on the subject give some idea of the global interest in the confluence of circumcision and HIV prevention: "The relationship between male circumcision and HIV infection in African populations" (1989) [33]; "Male circumcision: an acceptable strategy for HIV prevention in Botswana" (2003) [40]; "Neonatal Circumcision: A Review of the World's Oldest and Most Controversial Operation" (2004) [37]; "Randomized, controlled intervention trial of male circumcision for reduction of HIV infection risk: The ANRS

1265 trial" (2004) [39]; "HIV-1 target cells in foreskins of African men with varying histories of sexually transmitted infections" (2006) [34]; "Male circumcision for HIV prevention in young men in Kisumu, Kenya: a randomized controlled trial" (2007) [36]; "Male circumcision, religion, and infectious diseases: an ecologic analysis of 118 developing countries" (2006) [38]; "The potential impact of male circumcision on HIV in sub-Saharan Africa" (2006) [35]; and finally, an article to which we contributed in 2008, "A Practice for All Seasons: Male Circumcision and the Prevention of HIV Transmission" [29].

Awareness of the global picture can help us attack the problem at local and national levels. On the global stage, HIV/AIDS has reared its ugly head repeatedly. Its Medusa-like influence poses herculean challenges that seem unstoppable. It shows no favoritism on grounds of race, economics, social status, or gender. We must always remember that behavioral patterns lie at the root of the epidemic, and yet we must also recognize that there are striking differences in prevalence rates among those with similar patterns of high risk behaviors. Male circumcision clearly exerts a positive influence in the reduction of HIV infection. As we (Professors Addanki, Pace, and Bagasra) have written elsewhere, by way of summary, male circumcision (MC) "is known to significantly reduce female-to-male HIV transmission through sex, which then decreases male-to-female transmission. Three recent randomized controlled studies from Africa have shown that circumcision offers a 60% to 70% protective effect against heterosexual acquisition of HIV" [29]. We explained that "[t]he protective effect of circumcision against HIV, known since the 1980s, has been confirmed by more than 30 studies before…three famous randomized controlled trials, which are the criterion standard of clinical research." We reported that there is a dramatic reduction] in HIV prevalence in those countries with a rate of circumcision that exceeds 80%. Moreover, we found what others have also discovered: "MC not only reduces HIV but also other sexually transmitted diseases (STDs)." The convincing nature of the evidence has led to endorsement of male circumcision by leading world health institutions, including the World Health Organization (WHO) and the National Institutes of Health (NIH) [29].

According to Brian J. Morris, newborn circumcision has been on the increase in the US. The typically high rate for Whites has remained high, and the rate for Blacks has increased: "The rates recorded in the north-east region were steady at 70%, while rates rose in the mid-west (80%) and South (70%). For the western region rates have been falling due to the influx of Hispanics (50% of all births, so diluting out the overall rate in California to 35%)." He reported that "overall the statistics show an increase in circumcision rates for Non-Hispanic Whites," and explained that "in the West individual hospital data have shown...the rate for non-Hispanic Whites is in fact 75-80%." An alarming finding, however, is that "for the next generation of Hispanics, [only] 29% of boys are circumcised" [31].

The answers to the troubling rates of infection for Hispanics lie, at least partially, in educational and religious instruction, in both private and public settings. As immigrant Hispanics are more fully integrated into America's educational institutions, one would expect higher rates of positive health practices, including circumcision. Muslim and Jewish circumcision rates are very high, for religious reasons, but the persuasion of other groups may need to focus more on matters of health. Education, both in the schools and in the media, plays a major role in this arena. Where cost plays a role in the decisions about the circumcision of newborns, public policy makers might consider whether it would be in the best economic interests of the public to subsidize circumcision expenses, or pay for them entirely. They may also want to consider whether free circumcision for males of any age would be wise, on both health, and economic grounds. Many Latino immigrants are illegal folks and suffer from poverty. They cannot afford to pay extra $1,000 for circumcision for their male newborns. A free circumcision supported by the state or federal health care system would make more sense!

There are ways to tailor educational efforts to specific groups. Surely, our advertisers and scholars could suggest promising methodologies. Since country of origin seems to make a difference, why not give additional special attention to the most at-risk groups? If language barriers increase the probability of becoming HIV seropositive, why not make English language education more accessible (for a variety of reasons) and promote more bilingual communication. These are daunting challenges, and yet improvements can surely be made. Other challenges involve the role that hegemonic male-female power relationships (including *machismo* [41]) have on seropositivity, and how disparities in healthcare relate to HIV to AIDS progression. The health implications of double-standard *machismo* infidelity among Mexican men is serious, as Hirsch et al. conclude in their analysis of HIV factors for rural women in Mexico: "Marriage presents the single greatest risk for HIV infection among women in rural Mexico" [42]. Because drug and alcohol abuse contribute to the health challenges Hispanics face, would not stepped-up anti-alcohol educational efforts be warranted? Are not stricter controls on advertising alcohol a means of fighting battles against cheaper HIV roots rather than wars against costly AIDS branches? Ultimately, Acquired Immune Deficiency Syndrome (AIDS) is curable only to the degree that human behavior is improved. Given the bleakness that darkens the contemporary Hispanic health horizon, the partial solution emphasized in this paper, the MC solution, seems all the more inviting. Male circumcision may, in fact, be a realistic "behavioral vaccine" that can no longer be overlooked.

Beneficial health practices often coincide with religiously motivated behaviors. The use of tobacco was prohibited or discouraged by some religions long before United States surgeons-general began to routinely denounce tobacco as both addictive and deadly. Likewise, religious opposition to alcohol use predated the persuasive scientific data about the dangers of this substance, which is now commonly recognized as a drug. Male circumcision, now commonly performed for health reasons, also has deep religious roots. The removal of the male foreskin is an ancient practice that is well documented in the Christian Bible (*circumcision*) [7], the Jewish Torah (Pentateuch) (*bris*) [43], and the Muslim Sunnah (*tahara*) [44]. This practice is by no means universal, and is not practiced by some major religious groups, including Hindus,Buddhists and Sikhs [45].

Because of historical rivalry and even conflict between some religious groups, the decision to avoid circumcision can be both a matter of adherence to the practices of their own group and the rejection of the practice of the rival group. Hindu-Muslim tension provides a case in point. Islam continues to preach circumcision as an inspired practice that goes back to the patriarch Abraham. Hinduism has no such doctrinal or behavioral link to either Abraham or circumcision. Yet for a Hindu to reject circumcision because it is a "Muslim practice" would, perhaps, be understandable, but not sustainable on logical grounds. Numerous physicians, and others, perform male circumcision across the world, and these medical procedures may or may not have anything to do with religious doctrine or tradition.

The debate over circumcision might be but one of many instances of how religions differ in their doctrines and traditions. The debate has moved beyond a mere matter of religion, however, as confirmed by the growing body of scientific evidence that links circumcision to lower rates of HIV and AIDS. Circumcision is good for the human body, and what is good for the human body, whether considered good or not by a particular religious group, is good for the body politic.

Circumcision has proven to be a remarkably effective deterrent to infection, so much so that it may be characterized as a vaccine by minor surgery. Even without considering sanitation benefits, penile and cervical cancer prevention (30: pp. 50-61), and other health benefits, male circumcision is so successful in

HIV prevention as to merit consideration by public policy makers at all governmental levels. It is important to note that once a circumcised male has become infected with HIV, he is as likely to transmit the virus to someone else as is an uncircumcised male. Females are protected by male circumcision, therefore, by the lower incidence of infection that circumcision affords [36, 178-180]. Medical procedures cost money, and poverty can and does present an obstacle to proper care; so does ignorance. These two obstacles can be challenging hurdles for persons of any economic class, but for the poor they may be high hurdles indeed. Public funding may well be needed to lower or remove all barriers to circumcision.

To pay for an adult circumcision may seem uninviting to a male that is unconvinced about the utility of circumcision for himself. Funding agencies, public and private, may do well to consider providing financial incentives to male adults who agree to be circumcised. At-birth circumcision is generally easiest, and cheapest. It also makes monitoring most feasible. An HIV-related "tale of two cities" makes a strong case for intervention via circumcision, especial early intervention.

Lozi and Luvale are African villages located in the same valley in southern African nation of Zambia. Less than a mile separates them from each other, and their lifestyle patterns and sizes were likewise close. A major difference, one with profound health implications, separated them: the men of Luvale were circumcised for religious reasons; the men of Lozi were not. Early in the 1980s, the HIV/AIDS epidemic threatened the area. Villager soon noticed that HIV infection rates were far higher in Lozi, the town with the uncircumcised men. One-fifth of the young adults in the town tested positive while their counterparts in Luvale the rate was only about one-third that rate (7%). A calculated projection was made that if something were not done, Lozi would lose a staggering three-fifths of its children to AIDS. In spite of local tradition and the anti-circumcision position village elders, parents from Lozi commenced taking their children elsewhere to have them circumcised [30: p. 36].

This pattern symbolizes what is taking place across the world: circumcised men, like the men of Luvale, are far safer when it comes to HIV/AIDS. What is true locally, as seen in the previous example, is also true globally. A comparison of nations where circumcision is common with those where it is not shows a remarkable difference in HIV infection rates. Circumcision is a matter of individual choice, and such choices are more likely when adults are educated about the benefits they, and their families, and communities can receive by choosing circumcision. The circumcision decision is best made by someone other than the recipient of the procedure; infants cannot make this choice but it should be made for them, just as the decision to provide proper nutrition is made for them. Because of the vast numbers of uncircumcised males, the ideal public policy approach would be to encourage, with governmental financial backing as needed, adult male circumcision, and infant circumcision as well. Were this to take place, even without improvement in public morality, HIV infection rates would plummet.

Considering the enormous monetary sums that have been spent searching for a still elusive AIDS vaccine, public and private planners would do well to consider providing major funding to educate and persuade the populations of the world to promote circumcision.

Thus far, we have focused on the advantages of circumcision for males, but the benefits to females are also substantial. HIV-free males do much to keep wives, and illegitimate partners free of infection as well. Cervical cancer rates are far lower for women when the men with whom they are intimate are circumcised. When performed by proper medical procedures, circumcision is basically a positive benefit with no negative side effects. It is, as we have argued elsewhere, a "practice for all seasons" [29]. At the levels of both home and planet, male circumcision would improve both male and female health across the earth. Now is the time to plan globally, and circumcise locally.

# chapter ten

# ALCOHOL AND AIDS

زدی ا (AIDS in Persian)

Twins
Alcohol, debauchery
Vomiting, betraying, weeping
Bitter roots, bitter branches
Death-mates

THE WORLD HEALTH ORGANIZATION (WHO) has warned of an orally-ingested substance that plagues global health, and causes more deaths annually than does AIDS or tuberculosis or violence. That substance is alcohol, and it causes almost four percent of all deaths on the planet. Male alcohol-related deaths are particularly pronounced: "Globally, 6.2% of all male deaths are attributable to alcohol, compared to 1.1% of female deaths" [46]. The percentage for males in the Russian Federation and countries near its borders is much higher, approximately one out of every five. Globally, for about one in 11 individuals of those who die between the ages of 15 and 29, the cause of death is related to the use of alcohol [46]. Although alcohol use caused more deaths than any of the other three global killers, it is not possible, as the WHO recognizes, to artificially isolate alcohol: "Harmful drinking is also a major avoidable risk factor for noncommunicable diseases, in particular cardiovascular diseases, cirrhosis of the liver, inflammation of pancreas and various cancers. It is also associated with various infectious diseases like HIV/AIDS, STDs and TB, as well as road traffic accidents, violence, homocides and suicides" [100]. Those who choose alcohol typically choose much more than a mere drink. They also often invite terrible personal distress, health problems, family grief, physical suffering, emotional pain, and premature death. As terrible as is the toll that alcohol inflicts on the health of those who choose to drink it, the drug alcohol also inflicts terrifying consequences on mind and spirit.

> "The harmful use of alcohol is one of the world's leading health risks. It is a causal factor in more than 60 major types of diseases and injuries and results in approximately 2.5 million deaths each year. If we take into consideration the beneficial impact of low risk alcohol use on morbidity and mortality in some diseases and in some population groups, the total number of deaths attributable to alcohol consumption was estimated to be 2.25 million in 2004 (WHO, 2009a). This accounts for more deaths than caused by HIV/AIDS or tuberculosis. Thus, 4% of all deaths worldwide are attributable to alcohol. The harmful use of alcohol is especially fatal for younger age groups and alcohol is the world's leading risk factor for death among males aged 15–59" [46].

Human choice has so much to do with human behavior, and the choice to consume alcohol is notorious for impairing future choices. Indeed, at times, it is the choice to end all choices. The propensity to increase alcoholic intake has been linked to higher incomes in India, South Africa, and other African and Asian countries with sizeable populations [101]. It is important to remember, however, that although a pay raise may make purchases more possible, it is not the money that makes the decisions; humans do. It is the increasing frequency of destructive decisions vis-à-vis alcohol that causes such toxic pairings as alcohol-AIDS, alcohol-tuberculosis, and alcohol-accidents. Still, public alcohol policies commonly have not been elevated to a high-priority position on public policy agendas, even as elevated blood alcohol, accident, TB, and HIV levels increasingly clamor for their attention. And that is only the short list; the longer enumeration includes child neglect, poor job performance, lost employment, spouse abuse, unhappy relationships, violence, and the types of immoral behaviors that invite HIV into bloodstreams, homes, and communities. Public officials have the duty to balance individual freedom with the public welfare, and it is in no one's best interests to table serious policy discussions about alcohol/AIDS.

Scholars have long realized that a substantial number of HIV-infected persons drink alcohol. Although they continue to narrow in on the precise impact of alcohol on HIV proliferation, the consensus is that the news is bad. One study of Indian rhesus macaques found that after 18 to 24 weeks following infection, the group with alcohol had a plasma viral load that was between 31 and 85 times that of the non-alcohol control group. CD4 cell loss was significantly more pronounced in the alcohol group beginning as early as one week after HIV infection [205, 207]. Progression from HIV to AIDS comes much quicker when alcohol is involved [206]. The conclusion of the Bagasra team nearly two decades ago has been substantiated ever since: "HIV-1 replication may be augmented by alcohol in HIV-1-infected individuals, and alcohol intake may increase an individual's risk for acquiring HIV-1 infection" [208].

Educational officers at prominent US institutions of higher education have become alarmed at the collective impact of individual student decisions about alcohol. They are putting the matter higher on their already busy agendas. Ohio University, the University of Georgia, the University of Iowa, and the University of California Santa Barbara (ranked by the Princeton Review as numbers 1, 2, 4, and 5 among US "party schools") are making serious attempts to counter alcohol abuse, and its related problems, among their students. They, and others who deal with, or live among, persons who abuse alcohol (virtually everyone in the US), have abundant reason for alarm. Consider these statistics for the year 2009 from the "College Drinking—Changing the Culture" web site [102]. Death and injury: 1,825 college students, ages 18-24, died, and an additional 599,000 were unintentionally injured in cases where someone was under the influence of alcohol. Drunk driving: 3,360,000 students drove while under the influence of alcohol (little wonder insurance rates are high for young drivers) [103-105]. Academic problems: approximately 25 percent of students admit that drinking had a negative impact on their academic performance; it influenced their lower grades, missed classes, poor performance on papers and exams, and falling behind in their studies [106-109]. Assault: nearly seven-tenths of a million students were assaulted by a drinking student [103-105]. The list goes on, and includes sexual behaviors known to be directly related to the spread of HIV and AIDS: the number of alcohol-related cases in which 18-24-year-old students were victims to sexual assault or date rapes episodes were an astounding 97,000, but that number was dwarfed by the 400,000 students who had unprotected sex, and those troubled students (over a tenth of a million of them) who confessed to being so intoxicated that they could not say whether or not they consented to having sexual relations [102-105]. The probability of disease proliferation obviously soars on the drunken wings of such irrational behavior. Alcoholic cocktails can cause a desperate, life-saving need for retroviral cocktails. One leads to the other—the first cocktail by choice, the second a grasping-at-straws necessity. Yet the immediate cause of HIV infection has generally been the questionable, or blatantly immoral, behavior by one or both parties in an intimate relationship.

Immune deficiency is, of course, at the heart of AIDS-related deaths, and, according to the WHO, "alcohol consumption weakens the immune system, first reported by one of us in 1989 which increased HIV replication several fold (). In addition, the weakened immune system enables infections by pathogens, which cause pneumonia and tuberculosis. This effect is markedly more pronounced with heavy drinking and there may be a threshold effect" [46, 110]. For persons with immune systems already compromised by AIDS, alcohol-induced attacks on immunity are particularly troubling. Moreover, Baliunas, et al. [111] have reported what seems intuitively obvious: initial infection with HIV is more likely if an individual drinks alcohol. Hendershot, et al. (2009) [47] have observed that even HIV/AIDS patients who have access to antiretroviral treatments are less likely to benefit from those treatments if they drink because they are less likely to adhere to prescribed routines [46]. A BBC News play on word warned of the link between alcohol and immune deficiency illness: "Alcohol 'aids HIV cell infection'" [121]. Citing a study in the Journal of Acquired Immune Deficiency Syndrome, the BBC reported that alcohol predisposes cells in the epithelium (the lining inside the mouth) to be receptive to HIV infection. Following a ten-minute exposure to 4% of ethanol, HIV susceptibility was increased between three and six times. The deadly sequence, simply put, goes like this: HIV attacks CD4+ white blood cells. These infected cells then attach to the mouth's lining, to the cells of the endothelial. "HIV hijacks the cell, inserting its own genes into the cell's DNA and uses it to manufacture more virus particles. These go on to infect other cells" [121]. The process begins a chain reaction that can eventually produce full-blown AIDS, and then full-blown death. From both biological and behavioral perspectives, drinking and AIDS do not mix. Alcohol consumption increases the probability of HIV infection, and then weakens the immune system at a time when HIV-positive individuals desperately need healthy immune responses. Researchers are intrigued by the beneficial impact that the hepatitis-related virus GBV-Chas on HIV, but they find no such "co-infection" advantage with alcohol. To the contrary, alcohol kills, HIV kills, and, when combined, the two kill more effectively-a synergistic coordinated killing!.

To stay free from disease, the body must have a healthy immune system, but alcohol can damage routine immune reactions that provide necessary protection [51-54]. White blood cells (notably CD4+ T cells) are critical in the fight against infection, and while they do not fight HIV effectively, they do fight other infections very well. Not only is HIV not intimidated by CD4+ T cells, but it actually thrives on destroying them. This helps to explain why AIDS patients typically die of something not directly related to the HIV virus itself. The very name of the condition, immunodeficiency, is a key term in the acronym AIDS. Researchers have found that, in laboratory animals, alcohol intake causes a reduction in the total number of white blood cells that are available to counter disease-spreading pathogens [55-57]. Closer to home, they have found that the same holds true for humans [58-59]. Drinking alcohol can artificially limit the production of antibodies, and hold back other normal immune reactions in humans [60, 61], and animals [55-56]. Macrophages are a particularly helpful type of immune cell, and, among other things, they prevent lung infections. Like a villain attacking a hero, alcohol suppressed macrophage activity [62-63]. Alcohol consumption can even affect the next generation. Women who drink while pregnant can pass on weakened immune responses to their babies [64-66]. Indirectly, being high on alcohol can indirectly lead to being low on live-saving antibodies.

An intriguing study by Bagasra, et al. (1989) concluded that even one drinking experience could depress white-blood-cell immune responses. This investigation, based on research that used white blood cells from volunteers judged to be in good health, found that when isolated, these cells were more prone to infection by HIV retroviruses than were the cells of subjects who had not taken one or more drinks of alcohol [47].

Alcohol is a drug and, as with drug abuse generally, participants are more likely to use poor judgment, take unwise risks, and become infected with HIV. The National Institute on Drug Abuse (NIDA) explains, "drug abuse treatment is

HIV prevention" [48]. A mere glance at the titles of two recent scholarly articles by Baum, et al. suggests reason for concern, a reaction that is reinforced by the details in these articles: "Alcohol Use Accelerates HIV Disease Progression" [49]; and "Crack-cocaine use accelerates HIV disease progression in a cohort of HIV-positive drug users" [50]. Whether one argues that alcohol and drugs negatively affect behavior, and behavior in turn increases the chance of becoming HIV-positive; or whether one maintains that alcohol and drugs speed up the progression from HIV infection to critical immunodeficiency (AIDS), the conclusion is the same: alcohol, by itself, adversely affects human health; cocaine use, by itself, adversely affects human health; HIV adversely affects human health. Individually, they attack health; collectively, they assault it more frequently, and more quickly. Abstinence from drugs and alcohol is not only logical from an individual standpoint, but also from a public policy perspective. HIV statistics would improve, as would many other troubling indicators of societal wellbeing. It is time for the drunken wings on which disease proliferation soars to be clipped.

# chapter eleven

# HIV IN THE AGE OF THE *DESAPARECIDAS*

СИДА (AIDS in Macedonia)

**Gendercide**
**Illogical, contradictory**
**Testing, deciding, ending**
**Precious daughters left unborn**
**Imbalance**

WHAT DOES ABORTION HAVE to do with HIV infection? This question needs to be asked more often. The moral aspects of the question have been widely debated but what does abortion have to do with HIV? Abortion is commonly used as an attempt to erase the effects of selfish and irresponsible private behaviors, the same behaviors that vastly increase the probability of HIV infection. Human choices are at the root of both abortion and HIV infection. Even the innocent victims are, ultimately, affected by the unwise decision of some other human, however indirect that decision may be. As troubling as this is from the perspective of private morality and the rights of the unborn, the problems multiply when abortion is linked with gender selection.

Typically, there are 105 male births for every 100 female births. When this ratio is intact, populations persist, and the advantages of gender balance strengthen civilizations. Opposition is an inherent characteristic of human life, and this is apparent in the abuse of the marvelous capacity that ultrasound technology provides. Ultrasounds have done much to contribute to the health of both mothers and babies. When misused, as is commonly done in the world's two most populous countries, China and India, consequences flow that threaten the wellbeing of mothers, babies, and entire civilizations. Generally, ultrasounds can reveal the gender of a baby, but when mothers, or others involved persons, use that knowledge to promote the elimination of females, a deviation from the positive potential of technology occurs, and societies suffer [67-72].

China and India have the dubious distinction of leading the world in what has, with no exaggeration, been dubbed "gendercide" (a phrase coined in 1985 by Mary Ann Warren). In spite of weakly enforced Chinese and Indian prohibitions against using ultrasounds to unveil the gender of the unborn, Asian gendercide has become so prevalent that the losses in human life have become truly staggering. Surely, the intent of ultrasound inventors and developers was not to facilitate high-tech poaching against defenseless females. Journalist Mara Hvistendahl estimates that there are about 163 million Asian women who are not alive today who would be had it not been for a massive number of abortions. This number is comparable to that of all women now living in the United States. If the entire US female population were annihilated due to conscious human choice it would be deemed a disaster of the greatest magnitude, yet the fetus-by-fetus elimination of females in Asia receives only limited attention [67-72].

Ironically, a perverse manifestation of modern feminism has provided encouragement to the chief culprits, women themselves, in their determined war against female births. Their supposed private and collective victories are certainly pyrrhic, and the world is suffering, as it always does, from the effects of war. Nor is the problem strictly an Asian disaster. Azerbaijan, Armenia, Eastern Europe, and the United States (in particular population groups) have also contributed to this pyrrhic feminist victory. The global gender ratio has now been altered because of private decisions made in Asia, and elsewhere. Reviewers of Ms. Hvistendahl's *Unnatural Selection: Choosing Boys Over Girls, and the Consequences of a World Full of Men* (2011) [68] debate her interpretations, but they are justly impressed with her research, and the projected impact of her book. On a global stage with a preponderance of Asian actors, *Unnatural Selection* includes a varied cast that includes geneticists, mail-order brides, AIDS researchers, prostitutes, strident nationalists, fertility doctors, and prospective parents [67-72].

This number of missing females is far larger than the total number of AIDS deaths (perhaps five times as numerous). It is far larger than the number of *desaparecidos* (the disappeared ones of both genders, both *desaparecidos* and *desaparecidas*) that died in Argentina's "dirty war" of the 1970s and 1980s and in Iran during the Reza Shah's terror on his own people. The current attack on female life promotes an ever larger population of *desaparecidas*, a population notable for its absence rather than its presence. It is an absence that propels the probabilities of great civilizations in a negative direction, and such statistical trends are not simply due to extraordinary outliers, as can happen in calculations of wealth. It is individual decisions that have contributed to the vast throng of missing females. World War I is commonly blamed for significantly wiping out a generation of young men in France. Russian young men and those of other nations also suffered terribly. These *desaparecidos* caused incalculable heartache, altered demographic trends, and deprived their nations of children, inventions, writings, manpower, and so many other things. The total number of current *desaparecidas*, commonly carried out in the name of female liberty, may be about ten times the number of total deaths in the Great War [67-72].

Human decisions cause statistics; statistics do not automatically force human decisions. Still, it seems inescapable that the probabilities of violence, sexual promiscuity, HIV infection, and a veritable etcetera of other troubles will result from the *desaparecidas.* Asia is playing major leadership roles in the world today, but gendercide is not an area to be proud of.

Historically, the male-female ratio has been disrupted many times, but it also has a tendency to work its way back into the normal range. Unnatural events, such as wars, epidemics, and disasters of various kinds have affected the balance, but the pattern of selective abortions is unique to the modern world. In India, the ratio has risen to an abnormal 112. In China it is even more abnormal: 121, but this is just an average. The rate for many Chinese towns has skyrocketed beyond the 150 level. The abnormal ratio of the two Asian population giants is sufficient to skew the average for the entire planet to its present level of 107. Armenia's level is 120, Georgia's 118, and Azerbaijan 115. According to Ms. Hvistendahl, it is the rich who generally lead the way: "Sex selection typically starts with the urban, well-educated stratum of society" [68]. The journalist explains, "Elites are the first to gain access to a new technology, whether MRI scanners, smart phones—or ultrasound machines" [68]. She describes a filter-down process that institutionalizes abuse of technology downward to the poorer classes. She reports that the anti-female abortion trend is primarily promoted by the potential mothers themselves, and at times even the mothers-in law. In either case, it is a woman-versus-woman problem [67-72].

Ms. Hvistendahl warns that male-female imbalances create societies that can be miserable to live in: "Historically, societies in which men substantially outnumber women are not nice places to live" [68]. She warns of instability and violence, as occurred during the troubled times of Athens in the fourth century B.C., the Taiping Rebellion in China during the nineteenth century. In both cases, female infanticide was a troubling reality. The instability and violence in the frontier era of US history also gives evidence of the danger of letting male-female ratios get out of hand. In 1870, California had 166 males for every 100 females. The rate was even worse in Nevada (320), and incredibly out of balance in Kansas (768). In contemporary Asia, increases in crime have followed a rise in the gender gap. In India, income level is a leading indicator of the probability of criminal behavior, but the gender ratio is currently the best known predictor [67-72].

Several Chinese provinces have sex ratio at birth (SRB) levels above 130, including Hainan, in southern China, and Henan, a province in the north. India has normal SRB rates in the states of Andhra Pradesh and Kerala, but in the capital (Delhi), the northern breadbasket area of Punjab, and Gujarat, the SRBs are up to 125. Birth order is a statistically significant factor. Parents are less tolerant of female births as their number of children increases. Data from South Korea show how anti-female activity becomes more severe as male-childless couples have additional children. In 1990, the ratio in South Korea was a normal 104-100. Yet in that same year, the number rose to 113 for second births, soared to 185 for third births, and skyrocketed to 209 for fourth births. Moreover, immigrants to the US from China, India, and South Korea tend to enact birth selection activities in the New World that mimic the patterns of their ancestral homelands [67-72].

A look at the statistics for unmarried 28-49 year old Chinese suggests untold troubles for China, now and in the future. Only 6% of these single adults are female. Of the 94% who are males, few (only about 3%) have finished high school. Concerns about criminal activity, interpersonal violence, and psychological abnormalities seem well worth taking seriously. So many men will never have the chance for marriage. So much unnatural pressure will affect marriage decisions. So much will be out of balance because of the *desaparecidas*. Unmarried men are more likely, based on historical patterns, to have intimate relations with other men. This will increase the probability of HIV infection in Asia. Prostitution will be more likely, which likewise will increase the rate of infection. Crime will likely increase as well, as will the number of injecting drug users, a high risk group for developing, and dying of, AIDS [67-72].

What does abortion have to do with HIV infection? The

question needs to be asked more often. Nothing is not the answer. In the beginning, there was one Adam and one Eve. Things were in balance in those days, but today they are not in this age of ultrasound. Still, ultrasound is not the culprit; humans are. Much like the common pattern in which older immigrants persecute newer immigrants, too many of today's older females are given too many rights to persecute newer females. More Eves, and more Adams, need to write, persuade, speak, preach, legislate, and lead in a way that does not discriminate against women in ways that really matter, and in a way that does not allow the tyranny of some older females to rob newer females of their visas to places like China and India, and their passports to planet earth. What does such immigration policy have to do with HIV infection? A lot. HIV/AIDS is, finally, very much related to interpersonal relationships. Even with gender balance, HIV infection is a global pandemic. Gross gender imbalances will only make things worse. A well-known acronym, when applied to gender selection, might well be given double meaning: Acquired Immune Deficiency Syndrome, and Asia's Intolerant Daughter Siege [67-72].

# chapter twelve

# The Military-AIDS Complex: AIDS and International Security

에이즈 (AIDS in Korean)

| AIDS |
|---|
| Destabilizing, dangerous |
| Internationalizing, interconnecting, globalizing |
| Two complexes, too complicated |
| Casualties |

PART OF THE EXPANSIVE human dimension of the AIDS crisis in India relates to national and international security, including that of the United States. While good general health is important for the general populace, it is critical for military personnel. Very legitimate concerns over the alarming rates of HIV infection among troops in Africa have raised awareness of the need for healthy troops in India. Whether for civilians or military personnel, behavior that is both moral and safe is paramount. An 11 April 2004 CBS News report cited CIA concerns over AIDS involving military, economic and health considerations. India's independence from Al Qaeda in a region plagued with ties to the terrorism organization, plus its strong economic attachment to the United States are reasons to worry policy makers. The CIA report argued that a worsening of the AIDS epidemic in India would jeopardize its military strength by weakening the health of its army. AIDS would also hurt the South Asian nation's economy, which would no doubt send ripples through the American economy, with which it is closely linked. With one of six humans residing in India, and given the extensive travel in and out of India, an HIV infected India would likely lead to higher levels of infection elsewhere [122].

Over a year following al Qaeda's attack on New York's Twin Towers, Secretary of State Colin Powell urged the United Nations to recognize the link between AIDS and global stability:

"AIDS is more devastating than any terrorist attack, any conflict or any weapon of mass destruction. AIDS can destroy countries and destabilize entire regions" (cited by CBS, 11 April 2004). An important US ally in its drive against terrorism, India is an important American ally [122].

George Tenet, the director of the Central Intelligence Agency, argued that the HIV was a threat to national security that "can diminish military preparedness and further weaken beleaguered states." The security risk is linked not only to private moral behavior, which obviously has enormous implications vis-à-vis the spread of AIDS, but to economics. Antiretroviral drugs, which slow the progress of the disease, must be taken consistently or drug resistant varieties develop more quickly. Impoverished nations have contributed less to this problem for the tragic reason that their people often have no access to the drugs. India, more prosperous than the beleaguered nations of Africa, has the dual benefit of greater ability to purchase the medications and cheaper prices, due to lax patent laws. This easier access, however, is a two-edged sword: it prolongs lives but can lead to global drug resistance problem, particularly when patients do not, or cannot, take the medicine as prescribed. India's plight becomes the world's plight, given the interconnectedness of the planet [122].

Dr. Suniti Solomon, who in 1986 first discovered a person infected with the AIDS virus in India, remembered, in 2004, "I used to see one patient, new patient every week in 1991, 1992" and noted the alarming spread of the disease: "Today, we see 10 to 11 new patients every day." She established one of the earliest hospitals to deal with AIDS cases and, unlike other hospitals, sought to treat rather than avoid AIDS patients: "We don't throw them out." According to her, about one in five of the patients she deals with are truckers [122].

In the case of HIV/AIDS, lack of understanding can be deadly: "some truckers think bathing after being with a prostitute will do the trick, and they then go on their way making deliveries across this huge country -- delivering, among other things, the HIV to other prostitutes, and often to their own wives." Nine of ten of Dr. Solomon's female patients were infected by their husbands, to whom they were faithful. Solomon lamented the lack of protection available to women who were part of arranged, male dominated marriages [122].

In a complex and interrelated world, the military activities of the many nations influence each other. They also influence the political issues that governments deal with. Political and military decisions receive, and deserve, a good deal of attention, but so do health-related concerns. As noted earlier, the health status of India's troops influences the US. This is simply one brief case study from one large nation. The US, and other nations across the globe, have a vested interested in the health of the military forces of other nations. This may seem counterintuitive or even illogical. An outlook governed by malice and envy might wish for epidemics that would undermine foreign troops. While lacking in goodwill, an attribute much needed in the diplomacy of our era, such an outlook is unwise on both moral and practical grounds. Military personal interact with civilians, and both military and civilian individuals interact with others outside their boundaries. Good health for one helps preserve good health for all.

STI disease figures are typically higher for military personnel than for civilians. A statement from UNAIDS is cause for concern: "During peacetime, STI (sexually transmitted infection) rates among armed forces are generally two to five times higher than in comparable civilian populations; in times of conflicts, they can be more than 50 times higher" [123]. This is one of numerous reasons why war should be avoided as an instrument of international policy, except where clearly warranted on moral grounds. Most soldiers are fairly young, and are in a high risk age group. The use of children in the military, as is common in Africa, increases the probability of abuse, and of STI diseases, including AIDS. It was hardly encouraging, or inspiring, when The South African National Defense Force announced that rates among its military personnel were only slightly higher than the rate for civilian personnel. With adult civilian infection rates between 15% and 30%, the willingness to disclose infection rates may be commendable, but the infection rates certainly are not. Among the factors that increase the probability are a culture that promotes machismo and risk-taking, easy access to drugs and alcohol, widespread immorality, and fails to discipline senior officials who exploit younger troops, especially children [123].

According to figures for South Africa in 2002, AIDS-related deaths account for a stunning seven in ten (70%) military deaths [123]. "Uganda's defence force lost more soldiers to AIDS than to fighting in two decades of war with the Lord's Resistance Army. In Zambia, AIDS-related illnesses have killed more military personnel since 1983 than died in all its military operations combined, including the bloody independence struggle" [123-124].

In "Lessons learned," the concluding chapter of *The Enemy Within,* book editor Martin Rupiya explains the need for a complex attack on a complex dilemma: "There is need for a close civil–military approach to tackling the complex HIV/AIDS challenges. The impact of HIV/AIDS on the armed forces has demonstrated the old adage that no sector survives in a vacuum. Owing to the intimate linkages of prevention, transmission, care and treatment, counselling, nutritional support, home-based care, spouses and other partners, no approach will succeed unless it embraces contributions from, and collaboration between, civil society and the military" [124].

AIDS has proven to have great power to destabilize, especially when humans make unstable choices. Destabilization of individuals, couples, families, and nations. Unfortunately, that is part of the legacy of acquired immune deficiency syndrome, and its spinoff: the military-AIDS complex.

# chapter thirteen

## Counting the Costs: More Deadly than Military Conflict

Abivahendid (AIDS in Estonian)

| Massacre |
|---|
| Belligerent, vicious |
| Fighting, suffering, expiring |
| Large corps, abundant corpses |
| HIV |

NECROMETRICS MEASURE THE NUMBER of dead. Such numbers apply to the numbers of those who die in wars and from diseases. The American Civil War resulted in 620,000 deaths [21] the Spanish Civil War 365,000 [22] World War I 21,500,000, and World War II about 50,000,000 [23]. Each war caused suffering of troops and civilians. It was primarily caused by men, and fought by men. Men consciously killed men. Men purposefully imprisoned men. AIDS, likewise, kills. It wages war against humanity, and the number of deaths it causes may eventually eclipse the number of deaths in all the previously mentioned wars. The AIDS war may have seen only its beginning battles. Lustful men, in the name of love, inadvertently infect, and self-centeredly slay their male counterparts. They also ignorantly, and sometimes knowingly, infect their wives or female partners. In the early AIDS battles to date, women have also innocently, or guiltily, infected men and newborn children. Genuine love has sometimes facilitated the sowing of death's seeds in wives. Infected mothers, if known precautions are not taken, pass on more than a genetic heritage, and can simultaneously grant life and introduce death.

The American Civil War, Spanish Civil War, and the two World Wars of the twentieth century each lasted for a few years, but the war against HIV/AIDS has lasted longer than any of these military conflicts. To paraphrase Sir Winston Churchill, this ongoing war has not reached the beginning of the end, and we cannot even say with certainly that it has reached the end of the beginning [24]. Yet we see no dramatic change in human behavior, and no end to the deadly male-male relationships that launched the early *blitzkriegs* (German for lightning warfare). Moreover, sensually out-of-control tyrants selfishly spread the war's suffering to women. In search of lustful *lebensraum* (German for living space, a term used by Adolf Hitler), these morally-challenged Mussolinis selfishly strut onto the stage of public health.

Who-cares apathy breeds appeasement, even as the AIDS *Anschluss* (German takeover of Austria, without resistance) marches on. Frankly, Franco-like tirades against unnatural lifestyles [27] are excessively extreme and unreasonably lethal. Anti-HIV rhetoric must convert hearts and minds to change behavior, and sometimes partners. More than rolling-thunder [25] funding is needed to bombard HIV. Where the bombs are dropped does matter. Nor is any Hiroshima-Nagasaki [26] left-right knockout punch available. Given the absence of a preventive vaccine and the persistent presence of irrational human behavior, the fight between humans and AIDS will go on for many rounds.

## chapter fourteen

# "This is Not Your Land. Go Away AIDS": Music and Orphans

エイズ (AIDS in Japanese)

Children
Mommy, Daddy
Marry, infect, perish
Broken vows, lonely children
Orphans

AS A SERIOUS SOCIAL issue that is global in its impact and yet eminently local and personal, AIDS has inspired music that urges broader perspective regarding a major problem, and greater compassion toward individuals and families. As AIDS extends both its macro and micro reach, music provides one means of helping victims cope, and onlookers understand. Such understanding is much needed to encourage more caring interaction toward HIV-positive individuals who can safely interact with others in society, with no threat to others. Infected persons can do so much more if the stigma associated with AIDS can be weakened or removed. Jony Jerusalem, herself HIV-positive, wrote "The Aids song, The Stigma song" in response to her personal observations and feelings: "There is nothing I want more fiercely," she explained, "than to fight against and shatter the AIDS social stigma." She turned to music to publicize her thoughts and feelings:

> There are viruses, illnesses, cancers, disease
> That can wrestle the strongest of men to his knees
> But of all the diseases that cut short one's life
> Only AIDS has a stigma that cuts like a knife
>
> No, don't turn away from him, don't try to hide
> From the person who suffers this sickness inside
> No, don't turn away for the sake of a name
> For you know, deep inside, we are all just the same [74].

Barbara Luke writes of the visit that she and her husband Larry made to Uganda, and how they met with children who not only were HIV-positive but also had lost their parents. The Lukes are the United States managers for an organization known as the Child2Youth Foundation, an organization that supports HIV-afflicted children in various places of the African continent, provides uniforms to wear to school, furnishes educational supplies, and instills hope. With appropriate nutrition, financial sponsorship, and medical assistance, the lives of these children can more nearly parallel those of other children, and their lives can even be extended, perhaps for two decades or more. As the Lukes visited the schools that these "AIDS orphans" attended, they were welcomed with song and dance, but were surprised at the nature of this entertainment. Rather than hearing popular and cheerful songs about traditional childhood themes, they heard tunes that focused on the infection that threatened their own lives and dreams. Knowing the importance of children properly caring for themselves, and of their need to cope, teachers use music to teach these things. One group sings, "What have we done to deserve you, AIDS?" The tragedy of their query is deepened by their supporting musical declarations: "You are killing the rich, the poor, the old, all our families and, most of all, the young—the future generation—leaving too many orphans. This is not your land. Go away, AIDS" [75].

The Lukes observed other Ugandan children who intoned a morbid song over a little child they had symbolically wrapped in a blanket. The children began to weep; so did the Lukes. Their parents would weep as well, but they have departed into another sphere. Like actors forced to play roles in a tragedy they find objectionable, these dramatic players must contribute to a deadly plot they wish were nonexistent. Mrs. Luke echoed the feelings of so many as she searched for words: "What do you say to little children who feel abandoned and dare not dream of a future?" She assured them of unseen realities: of God, and of parents who live on elsewhere. She reminded them that there are visible beings that also care: "God sent us to tell you how perfectly beautiful you are, and we love you, too." [75].

As would be expected, AIDS orphans commonly experience psychological trauma. A study conducted in rural Uganda found psychological distress levels to be high. AIDS orphans were more likely to become angry, feel depressed, and suffer from anxiety. Some 3% of all children involved in the study said they wished they were no longer living; for AIDS orphans, the figure was four times that high. This study indicates that providing physical resources, which are too often lacking, is not enough. Serious social and psychological issues must also be addressed [76].

Psychological stress increases with separation from parents, but when this is coupled with separation from siblings, problems increase further. This is a regular occurrence in some regions. In Zambia, a 2002 survey found that over half of that nation's orphans did not reside with all of their brothers and sisters [77]. Orphans are less likely to have adequate food, shelter, clothing, educational opportunities, and healthcare. They are more likely to be part of a household presided over by a female, and such households are disproportionately large. They have more people but fewer wage earners to meet expenses [78]. AIDS orphans are often pressured to make financial contributions to those they live with, which increases the likelihood that they will turn to the streets for food or work. They are more likely to become beggars [79], which has its own risks and impacts on feelings of self-worth and personal safety.

Stigmatization continues to be a very real problem. Other children and even adults are not always understanding, and may act out of fear, ignorance, or superstition. Besides serious effects on orphans in terms of fear, anxiety, and rejection, these children also may lose access to common societal privileges. The death of a parent is sometimes accompanied by loss of property rights or denial of legitimate claims to an inheritance. As Pelonomi Letshwiti, a social worker in Botswana working for Childline, explained, "You find that the parents have been productive and have left assets for the children but immediately after their deaths, the relatives squander everything. Those that are left without anything are just being used for the food rations" [80].

Although illogical, AIDS orphans are often judged to be HIV-positive because one or both parents died of AIDS. Irrationality often breeds additional irrationality, and if children are deemed untreatable because of their presumed HIV infection status, then they may be denied medical assistance that should be rightfully theirs [81]. The parallel rise in the number of AIDS orphans and the growing number of young people involved in child labor is especially troubling [82].

Even with compassion and willingness to help in place, AIDS commonly imposes enormous burdens on families and communities. For African nations that have already suffered from civil strife, natural disasters, or epidemics, dealing with the AIDS pandemic is particularly acute. Caring for orphans is not a new issue in Sub-Saharan Africa. Now, however, families and community institutions have found themselves simply overwhelmed by the burdens AIDS inflicts, including lost income and higher expenses, such as those for medical care and funeral services [83]. As generous as donors have been to establish orphanages to care for the parentless children, these institutions should be regarded as only a stopgap measure, a temporary remedy. Life with other family members, or in foster-care settings, is generally preferable. Families are, after all, the basic institution of civilization, and efforts to approximate the family, where it does not exist fully, are generally better than placement in orphanages [81].

Uganda has an estimated 1,200,000 AIDS orphans. This is a startling figure, but Uganda has fewer parentless children than do Tanzania (1,300,000), South Africa (1,900,000), and Nigeria (2,500,000). Most of Uganda's orphans are not parentless because of AIDS, but an estimated 44% are orphans because of Acquired Immune Deficiency Syndrome. As terrible as this percentage is, it is lower than that of Kenya, whose 1,200,000 orphans comprise 46% of all orphans. The percentages rise even higher in Zambia (53%), South Africa (56%), Malawi (65%), Lesotho (65%), Swaziland (69%), Zimbabwe (71%), and Botswana (72). Staggering (Uganda, Kenya, Zambia, South Africa); more staggering (Malawi, Lesotho); most staggering (Swaziland, Zimbabwe, Botswana). [81]. The staggering macro trends imposed by AIDS are faced daily at the micro level of the individual AIDS orphan. The statement of a thirteen-year-old reveals the need for both material resources and human warmth: "My sister is six years old. There are no grown-ups living with us. I need a bathroom tap and clothes and shoes. And water also, inside the house. But especially, somebody to tuck me and my sister in at night-time." [81].

# chapter fifteen

# AIDS AND THE HISPANIC COMMUNITY

SEIF (AIDS in Irish)

Burials
Prolific, pathetic
Insert, close, bury
*Amigos*, *hermanos* left behind
Caskets

HISPANICS/LATINOS IN THE UNITED States face serious problems due to the HIV/AIDS pandemic. Moreover, the experience of this group, whose members can come from any race, has regional and global implications as well. Although in 2006 they comprised only 15% of the US population, Hispanics accounted for 17% of new HIV infections (Centers for Disease Control, 2006). By 2009, they accounted for 16% of the population but their HIV-infection rate had become even more disproportionate, having risen to 20% (9,400) of all new cases in the US. [84]. By 2009 they experienced triple the new infection rate experienced by whites (26.4 to 9.1 per 100,000), and the problem had become so severe that among those in the 35-44 age bracket it was the fourth leading cause of death. Obviously, this is a matter of major public policy concern. Current US immigration laws and regulations make living in families exceptionally challenging for Latino men who have come from Mexico, and other Latin

American countries. Stringent anti-illegal-immigration laws, notably those recently enacted in Arizona and Alabama, have a strong anti-Mexican and anti-Hispanic tone to them. Such laws reduce the opportunities for quality education among Hispanics, make self-sufficiency challenging, and threaten families. This may account, in part, for the atypically high percentage of men living with HIV/AIDS who contracted the disease through sexual contact with other men. While heterosexual contact accounts for most new infections in the US, and in the world generally, such is not the case with US Hispanic men. For this group, the three leading means of transmission are men who have sex with men (MSM), injecting drug use (IDU), and heterosexual relations with high-risk partners. For Hispanic women in the US, the two leading modes of transmission are high-risk heterosexual contact, and IDU. In 2009, 81% of the 6,000 new HIV infections among Latino males were from these MSM [73, 84].

The imbalanced male-female demographic patterns in many areas only partially explain the exceptionally high rates of new HIV infections among Latino men. Another cause, typically overlooked, is the lack of circumcision among this group. There is now overwhelming evidence that male circumcision dramatically reduces the possibility of contracting HIV. Unfortunately, the already low circumcision rate for Hispanic males seems to be sliding to even lower levels. This unhealthful behavior threatens not only the lives of the men involved, but also those of their spouses, children, and others [73].

A major category that the Centers for Disease Control (CDC) tracks that relates to AIDS cases among Hispanics is the country of birth for those diagnosed with AIDS in the United States. In 2006, the CDC reported that most cases of Hispanic AIDS transmission (2,608) did not come from a predominantly Hispanic country at all, but rather from the United States itself. Nearly half these 2,608 cases came from persons born in Puerto Rico (1,346), which was very close to the number born in Mexico (1,334). There were 814 diagnosed cases for natives of Central and

South America, and 145 for those born in Cuba. In each of these groups, except Puerto Rico, most cases were attributed to "Male-to-male sexual contact," followed by "High-risk heterosexual contact," and then "Injection drug use." Most of the Puerto Rico-born cases were the result of "Injection drug use," followed by "High-risk heterosexual contact," and then "Male-to-male sexual contact. [86]

US AIDS Transmission by Country of Birth

| | Central / South America (N = 814) | Cuba (N = 145) | Mexico (N = 1,334) | Puerto Rico (N = 1,346) | United States (N = 2,608) |
|---|---|---|---|---|---|
| Male-to-male sexual contact | 49% | 62% | 59% | 18% | 45% |
| High-risk heterosexual contact | 36% | 24% | 26% | 36% | 25% |
| Injection drug use | 10% | 9% | 10% | 40% | 22% |
| Male-to-male sexual contact and injection drug use | 3% | 4% | 4% | 5% | 6% |
| Other* | 1% | 0% | 1% | 1% | 2% |

*Includes hemophilia, blood transfusion, perinatal exposure, and other risk factors not reported or identified. [Adapted from source 86].

The types of behaviors that lead to HIV infection are typically self-defeating and irrational. Alcohol abuse, for example, markedly increases the likelihood of unwise decisions regarding intimacy, which in turn take a terrible toll in terms of HIV/AIDS. Lack of condom use, unsafe activity with multiple partners, refusal to abide by high standards of moral behavior, and failure to get tested are likewise irrational *human choices*. More than self-defeating, they are often other-defeating as well. It is important that HIV seropositivity does not become a contemporary rite of passage for young Hispanics who engage in high-risk behaviors" [73].

AIDS transmission is, of course, linked to other social and behavioral factors. According to the CDC, "Surveillance data show higher rates of reported STDs among some minority racial or ethnic groups when compared with rates among whites." In the US, race and ethnicity clearly correlate with what the CDC calls "more fundamental determinants of health status." Poverty certainly exerts its impact, as do the related areas of access to healthcare, quality of healthcare, social atmosphere, drug availability, and behavioral modeling. Admitting the severity of STD rates and acknowledging their link to issues of ethnicity and race are important in coming to grips with solutions for prevention ("STDs in Minorities"). Regardless of factors that increase the likelihood of infection, however, it is crucial to remember that these factors to not make decisions; humans do. Poor Hispanics can choose to act rationally regarding private moral behavior. Hispanics whose parents have provided poor models of intimate behavior and alcohol use can choose to depart from these self-destructive models [73].

Young Hispanic males are particularly at risk to make dangerous choices. Of the new HIV cases in 2006 among Hispanic males, 41% were among those 13-29 years of age, 16% percent higher than the corresponding figures for White, non-Hispanic males (25%) in that category, but 1% lower than the 42% registered by Black, non-Hispanic males. Hispanic young women in that age range comprised 36% of the total, compared to 32% for both White, non-Hispanic; and Black, non-Hispanic females. In the 30-39 year age range, the percentage of new HIV infections for both Hispanic males, and White non-Hispanic males stood at 34%, a figure slightly higher than the 31% for Hispanic females, and 32% for White non-Hispanic females. For Black, non-Hispanics, the figures were 26% and 30% for males and females respectively [73].

On a percentage basis, the aggregated figures for 13-39 year-olds were most bleak for Hispanic males (75% of the total Hispanic male cases), followed by Black non-Hispanic males (68%), Hispanic females (67%), and Black non-Hispanic females (62%). On the basis of total numbers (as opposed to percentages), most 13-39 year-old seropositive male infections came from Black non-Hispanics (10,930), White non-Hispanics (9,650), and then Hispanics (5,530). The order was the same for 13-39 year-old females: Black non-Hispanic females (5,480), White non-Hispanic females (2,110), and Hispanic females (1,540) [73].

For new HIV infections among those in the 40 and older bracket, most cases combined came from Blacks (8,510), followed by Whites (7,820), and then Hispanics (2,650). However, it was White males (6,620) who had the most cases of any of the gender-specific groups. They were followed by Black males (5,190) and Black females (3,320). White females accounted for 1,200 new cases, Hispanic males 1,890, and Hispanic females 760. In this older bracket, Black females (3,320) had more new infections than Hispanic males and females combined (2,650). ("Subpopulation Estimates from the HIV Incidence Surveillance System --- United States, 2006") [29, 73, 85-99]

# chapter sixteen

## CULTURE AND AIDS TRANSMISSION: THE EXAMPLE OF INDIA

एड्स (AIDS in Hindi)

Prevalence
Disloyal, risky
Driving, piercing, not-circumcising
Known perils, menacing future
Achilles-heels

ANNUALLY, THE NATIONAL AIDS Control Organisation (NACO) and other institutions give estimates regarding HIV/AIDS statistics for India, a country in which over one billion people now live. These latest results, for 2006, include a major adjustment of an earlier exaggerated estimate of 5.2 million people living with HIV/AIDS. The estimated prevalence is about 0.36 percent, or between 2.0 and 3.1 million individuals. NACO uses an average figure of roughly 2.5 million. While that is only about 50% of the earlier estimate of 5.2 million, it is still serious [112]. India's prevalence rate is not nearly as high as some African countries, such as South Africa, which had a 2006 prevalence rate of about 29.1% (down to about 28.0% by 2007) [113]. Nevertheless, due to India's vast population, it is critical to prevent India from ever reaching a threshold from which it would be difficult to prevent a rapid upward spiral of new cases. We have traveled multiple times to India, have made presentations about HIV/AIDS, and have come to some conclusions based on our research, conversations, and observations. These conclusions are very much culturally based. HIV infection is integrally related to cultural and behavioral patterns. India, with its exceptional cultural diversity, offers an excellent case in point.

HIV/AIDS is not only a biological and medical problem, but also a serious communication challenge made more difficult by the complex cultural patterns of India. Geography, linguistics, religion, and private behavior are all closely linked to culture. India's cultural diversity is both enriching and challenging. Only a culturally-sensitive approach to fighting HIV/AIDS has the potential for cutting through the cultural complexities in a way that is understood by the Indian population. Indian culture often mitigates against coordinated intervention by governmental, non-governmental organizations (NGOs), and educational institutions, all of which are needed in tackling the interrelated challenges posed by those most responsible for spreading the HIV virus, including truck drivers (noted for their promiscuity), female sex workers (FSW; an obvious high-risk group), men who have sex with men (MSM; easily the highest risk group in India), and intravenous drug users (IDU; infected through unsanitary use of needles). Categories help provide focus, but there is subjective decision making beneath the apparently objective statistical summaries. For example, of the 2006 total of adult AIDS infections in India, 86.2% came from the general population, and the rest were from four high risk groups: MSM (6.7%), truckers (3.6%), FSW (2.8%), and IDU (0.7%). What if additional high-risk categories could be added, such as those who act under the influence of alcohol were a risk group, those with no strong affiliation with a religious institution, those who live in poverty, those who have intimate relations outside of marriage, or those with very limited formal education? Granted that data collection for these categories would be challenging, but the point we make is that the 86.2% figure for the general population may mask more than it clarifies [112].

There are several so-called Achilles' heels that may be key culprits in the spread of HIV in India. These include: a truck drivers' culture that encourages high-risk and immoral behavior while on the road, the mistreatment of widows in Hindu culture, the Devadasi system, the absence of circumcision in both Hindu and Sikh cultures, the widespread use of improperly sterilized needles in hospitals, clinics and physicians' offices, the nature of HIV/AIDS prevention efforts by the Indian government and NGOs, and systemic mistreatment of women in ways that spreads the epidemic. Our travel, contacts, and research have enriched our knowledge about cultural, social, economic, and religious customs in India, and suggested potentially successful preventive strategies that can stymie the spread of HIV.

In India, the most common source of material goods transport and delivery is small trucks. Still, India lacks the infrastructure of the US and other western nations in which large superhighways accommodate transportation of food and material supplies at high speeds. Small trucks are very efficient in carrying out this function, but often are not allowed to travel in the daytime because of heavy traffic and relatively small roads that are mostly unpaved. While this no doubt reduces daytime vehicular accidents, it indirectly facilitates greater promiscuity, and the accompanying spread of HIV/AIDS. For religious reasons, cows and bulls are commonly allowed to block the small roads that connect India's villages, and even the congested roads of the cities. Truck drivers generally spend their days in truck stops (dhabas) that generate significant income for local economies. These are not the kind of truck stops we see in the US, but are small, makeshift sites where drivers stop in the daytime. Truckers find small shops that provide barbers, truck repair personnel, food vendors, and other services. Unfortunately, these stops also have sex workers who exchange their services for money. These are not professional sex workers, but often farm workers who carry out this role to supplement their incomes during the off-season or other times when off work.

There is now clear evidence that male circumcision (MC) significantly reduces female-to-male transmission of HIV. (We also maintain that MC reduces male-to-female transmission.) In the first randomized controlled trial (RCT) to report on MC, Auvert and colleagues [39] have shown that MC reduces transmission from women to men by 60% (32%-76%; unless otherwise stated, ranges are 95% CI). In sub-Saharan Africa, estimates of HIV prevalence are significantly associated with the estimated prevalence of MC. In countries where fewer than 30% of men are circumcised, the median prevalence of HIV is 17%; where more than 90% of men are circumcised, it is only 2.9%. We maintain that the Indian masses must be convinced, regardless of religious perspective, that it would be greatly beneficial for them to practice circumcision. Our preliminary data show that Africa, with high rates of HIV/AIDS prevalence, still had varying infection rates, with the lowest rates in the areas where a majority of the population practice circumcision (Muslims), whereas the highest number infected was in the uncircumcised groups. India echoes this pattern, with the majority of its infected population coming from the southern region, an area with a low percentage of Muslims. Regions with high percentages of Muslims were observed to have less than 0.1% of their population infected with HIV [115].

Clean needles are troublingly unavailable in India. Particularly disturbing is the pattern by which low-level staff members hoard clean needles, only to sell them on the black market. Even HIV/AIDS clinicians use reusable needles that are "sterilized" in warm water. This appears to be a widespread practice throughout India. In addition, tattooing and shaving by barbers with unclean blades are common practices in India that can contribute to HIV transmission. We maintain that the widespread use of relatively contaminated needles for basic medical procedures is one of the major culprits in the spread of HIV. These include the use of such unsafe needles even by the staff members of physicians and nurses, who generally are uneducated in sterilization techniques [118]. We propose that Indians should be educated in this urgent matter by mass media advertisements. Since a significant percentage of Indians are uneducated they need to receive information by other means than reading brochures or handouts (preferably by TV). This will require actual visual images from TV and movie previews. In addition, people need to be educated about other potential means of getting infected than through sexual activity, including tattooing, body piercing, and intravenous drug use. India's National AIDS Control Organisation (NACO) places tattoos among its HIV risk factors. It warns against "Injections, tattoos, ear piercing or body piercing using non-sterile instruments" [119]. The Centers for Disease Control in the US also warns about the potential of becoming AIDS infected through tattooing:

> A risk of HIV transmission does exist if instruments contaminated with blood are either not sterilized or disinfected or are used inappropriately between clients. CDC recommends that single-use instruments intended to penetrate the skin be used once, then disposed of. Reusable instruments or devices that penetrate the skin and/or contact a client's blood should be thoroughly cleaned and sterilized between clients...
>
> Personal service workers who do tattooing or body piercing should be educated about how HIV is transmitted and take precautions to prevent transmission of HIV and other blood-borne infections in their settings.
>
> If you are considering getting a tattoo or having your body pierced, ask staff at the establishment what procedures

they use to prevent the spread of HIV and other blood-borne infections, such as the hepatitis B virus [9].

Obviously, the safest and cheapest course to take is to simply avoid all tattooing and body piercing. However, this hardly seems practical in contemporary India since virtually all Indian girls get their ears, and right side of their nose, pierced. Hundreds of millions of people are involved; sterilization of needles, therefore, is imperative.

The major efforts of Indian governmental authorities are directed towards a few states that appear to have the highest incidence of HIV infections. The alarm regarding these areas is understandable, but given the finite resources available to fight HIV/AIDS, care must be taken to balance prevention and treatment, and to balance the relative amounts of aid going to the various regions of India. HIV prevalence data for each state is primarily established through antenatal clinics, where pregnant women are tested. Although these data only directly reveal statistical information for sexually active women, they can have general interpretive value when intelligent data interpolation takes place. Much more needs to be done to *prevent* rather than *react* to existing problem areas. Greater advocacy by higher education (including students, who themselves comprise a high-risk group) and by NGOs (who need greater accountability [118, 120], motivation, and training materials) is needed. While these groups are making some genuine efforts, the HIV *crisis* requires an increased sense of urgency. Higher education has a unique, but unfulfilled, opportunity to play a leading role in the fight against HIV/AIDS. Administrators, faculty, and students hold an important key in the effective dissemination of accurate information to their local communities. With their linguistic expertise, social status, political clout, and communication potential, the higher education community has unparalleled potential to teach India what it must do to avert even more serious disaster.

Given the complexities of Indian culture, a multi-pronged response is needed to attack the multi-faceted crisis. Such a response must urge an attack on AIDS that is conducted in a culturally sensitive manner. Given the variety of cultures in the country, we might more accurately speak of *Indias* and of HIV/AIDS *crises*. Each of these *Indias* would benefit from a systematic, nation-wide approach to prevention and cure, but in the national and global war against HIV/AIDS, each battle must ultimately be fought on a separate, more localized, battleground by forces that are linguistically and culturally capable of confronting the epidemic in ways that are culturally-sensitive.

## chapter seventeen

# Indian Traditions, Women, and HIV

எய்ட்ஸ் (AIDS in Tamil)

Ashram
Widowed, abandoned
Grieving, coping, dying
Culturally entrenched superstition persecutes
Alone

WOMEN, AS WELL AS children (especially orphans), bear disproportionate burdens in contemporary India. Issues related to the exploitation of women also commonly have application to children. The female portion of the population has the least power to deal with the numerous issues that HIV/AIDS has inflicted upon them. The limitations imposed by economic hardship and prejudicial cultural views about women are not only demeaning to women but also have an impact upon the spread of disease. While much progress has been made to curtail the oppressive demands of the dowry tradition, a tradition upheld financially by the bride's parents, this unfortunate social convention continues to have an impact on marriage opportunities, and marital tension. The lack of a dowry, or low amount of a dowry, makes marriage less likely, and increases the likelihood for promiscuity, with its attendant health risks. Increased potential for education has resulted in improved conditions for many women, but still vast numbers of women are inadequately educated. Moreover, cultural taboos prevent women from having the voice that is needed in sounding the alarm and teaching others, including to other women, about the myths and realities of HIV/AIDS. We have observed that the large majority of those with whom we have dealt in higher education are males, while neither higher education nor society in general can fight HIV adequately without the serious and pervasive involvement of women. Given the subordinate role in which women and children find themselves in India, we argue that males must shoulder the primary responsibility (and accept most of the blame) for modeling moral and ethical behaviors that are compatible with the best of the various religious and social traditions in India, behaviors that lie at the heart of HIV/AIDS prevention.

In Hindu culture, once a woman from a lower cast becomes a widow, she is not allowed to remarry and often goes to live in an *ashram*, where she lives an ascetic life along with other widows. This inability to remarry, although not unique to Hinduism or to India, seems to have an adverse impact on public health. The residents of the *ashrams* commonly live in poor and destitute conditions for the rest of their lives, and many become targets of wealthy men who exploit their vulnerability for selfish immoral purposes. According to the 2000 census report, there were 3.4 million Indian widows living in *ashrams*, many near a small or a midsize city. India's Muslim population is the second largest in the world [114]. Islam allows widows to remarry, which alleviates many of the problems that widows face in India.

As discussed in one recent study, "Widowhood in India is a complex institution fraught with contradictions in meaning and practice." Particularly among the Hindu majority, widowhood isolates women from society, deprives them of normal social and economic privileges, increases potential for loneliness, decreases the likelihood of receiving proper health care, reduces their income, imposes social stigmas, blocks them from future marriage opportunities, and imposes restrictive expectations regarding food and clothing. "Perhaps more than any other social institution in India, widowhood exposes the gap between cultural and social realities, between precept and practice" [114]. Granted, widowhood poses challenges to the husbandless women, and to society, in the best of circumstances. But conditions in India are particularly perplexing, and widows are particularly vulnerable. The vast number of widows in rural areas find themselves incapable of managing crop and land issues without male help. Moreover, the medical debts associated with their now deceased husbands become their responsibilities. Unable to meet the rigors of rural life, they feel pressured to migrate to urban areas, and yet this migration comes with its own risks "of trafficking, physical and sexual assaults etc[.] and many young widows end up in the brothels" [114]. As with avaricious funeral home services in the US, where the next of kin of the deceased are pressured into paying exorbitant prices for caskets, and then left with thousands of dollars of loans for burials, there are specialized groups in India that lure the newly widowed into entering a new "life" full of promises and hopes (). In short, there is, in contemporary India, a troubling correlation between widowhood and AIDS. The conditions that challenge millions of Indian widows increase the probability of HIV infection in the world's second most populous country.

Devadasi is a religious practice in parts of southern India, including Andhra Pradesh and Hyderabad, whereby parents marry a daughter to a deity or a temple. The marriage usually occurs before the girl reaches puberty and requires her to provide sex to upper-caste community members. Whether this lifestyle is viewed as prostitution or religious zeal, it prevents these *jogini* from entering into a real marriage. Mythically, the Devadasis are the incarnation of Urvashi, the celestial nymph. *Joginis* are recognized by their copper bangles, the band they wear around their necks with a leather pendant, and a long necklace with several pendants which have the image of the goddess Yellamma. The practice was legal in much of India until 1988, yet there are still instances of the practice due to continuing superstition, slowness to accept societal reform, and the unwillingness of authorities to intervene. This type of religious prostitution is known as *basivi* in Karnataka, and *matangi* in Maharashtra. It is also known as *venkatasani, nailis, muralis* and *theradiyan.* This practice gives evidence of the link between religion, superstition, exploitation, and HIV transmission.

We recommend an aggressive program for the eradication of the *jogini* system, which is a remnant of the Devadasi. It is a heinous practice that is often thrust on the poor, "untouchable" women in the remote villages. Large-scale TV advertisements, as well as counseling programs, would help change the attitudes of the *joginis* as well as of the villagers. The enactment of the Andhra Pradesh Devadasi (Prohibition of Dedication) Act of 1988 is commendable:

> Whereas the practice of dedicating Women as Devadasis to Hindu dieties [deities], Idols, objects of worship, temples and other religious institutions or places of worship exists in certain parts of the State of Andhra Pradesh; and
>
> Whereas such practice, however ancient and pure in its origin, leads many of the women so dedicated to degradation and to evils like prostitution; and
>
> Whereas it is necessary to put an end to the practice [116]

The long-overdue legislative enactment in Andhra Pradesh was preceded four decades earlier by a similar measure in Tamil Nadu [117]. Both laws deal with the issue of marriage, and make it illegal to prevent a previous Devadasi from marriage: "any woman so dedicated ["to the service of a Hindu deity, idol, object of worship, temple or other religious institution"] shall not thereby be deemed to have become incapable of entering into a valid marriage" [117].

However commendable the wording of the 1947 and 1988 Acts for Tamil Nadu and Andhra Pradesh may be, they also need to be enforced on a continuing basis, widely publicized, and imitated in order to bring greater hope to impoverished poor women, to heap greater shame on those who use social and religious pretexts to prey on others, and to fight AIDS. While TV holds great promise for the dissemination of accurate and helpful information regarding HIV/AIDS, it also has much negative potential. Indian culture is increasingly influenced by television. Western influences too often encourage the very behaviors that breed disease. We maintain that government has the right and obligation to regulate the media where matters of public health are concerned. We strongly recommend that the religions of the world be used to further practices that ennoble both males and females, and promote the public good. Acting in accordance with harmful and uncivilized superstitions, regardless of how ancient or enshrined in practice, should not be allowed to exist legally in the civilizations of the modern world, including the great and ancient civilization of India.

# chapter eighteen

## An Insightful Interview with an Indian Physician

סדי״א (AIDS in Hebrew)

| Avarice |
|---|
| Incredible, unnerving |
| Observing, tolerating, whistle-blowing |
| Ethics less than knowledge |
| Scam |

THE CHAPTER FOLLOWING THIS one deals with the issue of unsafe use of needles in healthcare. Since it is based on an interview with a physician trained in India, the present chapter is given so that readers may learn from the interview directly, a primary source upon which the subsequent chapter is based. The interview between the authors (Dr. Omar Bagasra and Dr. Donald Gene Pace) with the physician (referred to as Dr. MM to preserve anonymity) took place on 12 February 2007. It was held in Dr. Pace's office on the campus of Claflin University in Orangeburg, South Carolina, United States of America. Born in the United Arab Emirates, Dr. MM studied medicine in India between 1998 and 2004, where he gained first-hand experience in observing the strengths and weaknesses of medical practices in contemporary India. The authors acknowledge that a single interview cannot describe the medical practices of an entire subcontinent; however, this type of primary source can provide an in-depth glance into a variety of practices and issues of medical and social importance.

Dr. Bagasra Tell how do you have connection to India?

Dr. MM My mother is Indian and my father is Arab. I have been brought up in mixed culture. I can speak Arabic and Urdu, and I understand Hindi. Many of my family friends are from either India or Pakistan. My grandfather served in the... army in Hyderabad in early 1940. His family had to migrate to Pakistan during the partition time of India- Pakistan because of the killing of Muslims and Hindus due to the misunderstanding between them. My father settled down in Karachi, completed his civil engineering degree, went back to India, and got married.

Dr. Bagasra We are interested in knowing about the HIV status in India. You have stayed in India during your medical school and have some idea of what is going on there...so please elaborate.

Dr. MM I was in India between 1998 and 2004 attending my medical school and I have been exposed to private and government medical facilities, in urban and rural areas. In my opinion, HIV is underestimated in India to great extend and the government does not want to talk about it nor make it a priority. I guess Indian government has a strange policy to ignore any crisis where it can shed some population, and that is why in most of the crisis their response is always late. This policy is made to control the growing population by some means. Indian government is behaving the same way about HIV and not taking the right measures, so people will die and only the fittest will survive.

Let us talk about HIV and how private medical facilities deal with it. Private medical settings are booming now in India because the life style in India is becoming better, lots of multinational companies with good pay and other benefits. People can afford private hospitals unlike before. In such place if they come to know the patient is HIV positive they will throw him or her out of the hospital except if the center in big and reputed. This kind of behavior is result of the fear that if other patients come to know that this clinic or hospital treats HIV patients and in future these patients will not come again to this hospital.

If the patient is due to [have] an interventional procedure or an operation the hospitals demand HIV testing, and in case of clinics they ask the patient to bring HIV result in the form of the real testing kit used on him and not just the report. Many of the medical labs in India report results without performing the real test so they save the money of the testing kit or reagent and make more money. This can be disastrous in case of HIV and that is why now medical centers want the real kit used and not the report; kit should be sealed with your name on it. Most of the centers now do the test in their own setting and in case the test is positive, the patient is asked to leave.

In government setting, similar things used to

happen. I was working in one of the biggest government hospital OBGYN departments and all ladies were asked to bring a test called three dot Elisa test from private labs. This test was not available in that hospital. Ladies was asked to get the test sealed with their names on it and if it is positive then nobody is going to help that patient.

Dr. Pace — In case of a woman who is giving birth, what was the reaction towards HIV positive women?

Dr. MM — If the lady have already entered labor then they can't do anything and will help her. However, if by any means they come to know this lady is HIV positive, nobody will even care to assist her in giving birth.

I remember the incident that happened to my brother's friend. My brother attended medical school in India in the same city as I did and he was two years senior to me. His friend was doing training in some government hospital. Some patient was found to be HIV positive, and in terminal condition. The doctors decided to leave him because they estimated he would not live more than 4-5 days. The patient was septic and needed blood to be taken for blood culture; however nobody was ready to prick him with needle and take blood for the test. My brother's friend volunteered to draw the blood and, unfortunately, he pricked himself while drawing the blood. He was tensed and lost control, so he hurt himself. This matter became a big matter in the hospital and the patient was thrown out of the hospital and died within 24 hours later.

So this shows how doctors are not ready even to draw blood to help HIV patients. I am not generalizing, there are some hospitals who will deal with HIV patients but charge heavily for the strict infection control procedures and few Indians can afford such expensive places.

Dr. Pace So poor people are not being treated if they are HIV positive?

Dr. MM Yes, poor people were not being treated until this new business started. Some labs for money, they will give you an HIV negative result and even fake and kit and the patient have to pay little extra money for this kind of crime. Once the patient gets this negative result they can be treated anywhere with nobody even knowing that this patient is HIV positive. This sort of practice makes this patient real threat and may aid the spread of the disease. This practice pushed many center to do the HIV test themselves.

This is what happens usually in big cities and towns. In rural areas, some other kind of problems usually encountered. Most of the Primary Health Centers (PHC) do not get enough medical supplies from the government but this is not the major problem. The problem is most of the supplies to these PHCs do not make it to there and are sold in black market!

Dr. Pace What do you mean?

| | |
|---|---|
| Dr. MM | I mean the following, PHCs get needles and drugs and other stuff as supplies from government and should be used and given for people in rural villages for free. This does not happen. I was working in the PHC in some remote area where we use to have two needles in dirty boiling water and these needles where used on 50 to 100 patients a day until it broke then new needle is opened. |
| Dr. Pace | Why so? |
| Dr. MM | Because there is no needles; if you ask for new needle they will tell you there is none! |
| Dr. Pace | These are medical professionals? |
| Dr. MM | Not usually; most of the doctors live in towns and cities, go to PHCs by car or bus in morning, and come back by evening. Person in charge is, most of the time, some 10th grade graduate who takes care of the building and the store. This person is from the same village and lives near the PHC. He sells all the free needles and drugs in black market to make money, and sometimes some doctors take share in it. If the needles are not sold in black market, it is sold in the same PHC to the patients; in case a patient demands clean needle they instruct him to buy it from them! |
| Dr. Bagasra | Some of this stuff is given by UN [United Nations] and other agencies but never reaches the people! |
| Dr. MM | They claim the water is boiling so the needles are clean but the problem is when the water is not boiling, which happens very often. The boilers |

used either run on gas or electricity, and in either case if gas is over or electricity went down–that happens very often in those areas–the water is not boiled and needles are still used.

Dr. Pace — But if the water is boiling, the needle is sterile?

Dr. MM — Not exactly, there is certain temperature, for certain amount of time, needed to consider the needle sterile, but if you are using the needle in one patient and after two minutes of boiling use it on the next patient, I don't think it is sterile; guess the temperature needed is 124 [degrees] for five minutes, not sure.

Dr. Pace — If you know that, they must know that too.

Dr. MM — They know it and, if you ask, they will answer, "What do you want me to do? Shall I get the needles from my home?"

Dr. Pace — Scandal!

Dr. MM — If you ask, "Where is the needles provided by the government?" the person in charge says the government did not give any needles this month or the last three months. Later on, you come to know that it is already sold in black market.

Dr. Pace — To whom are they selling these needles?

Dr. MM — Drug shops (pharmacies) outside the village, and sometimes to the patients attending the PHC if they want clean needles or drugs.

Dr. Pace — Making extra money!

Dr. Bagasra In India unlike US, you can go and buy drugs from pharmacy with no prescription; middle class people go and buy their own drugs and clean needles because they know about HIV. However, the majority of poor people do not even know about it and they think doctors are taking care of them!

Dr. MM Some of these persons in charge of PHC run their own drug stores; they have enough drugs and supplies to run a drug store!

Dr. Pace These people are not highly educated people.

Dr. Bagasra Fifth grade, sixth grade!

Dr. MM I remember, as a foreigner, I was instructed to have an HIV test within 40 days of entering India, and whenever I used to go to the center of testing, they used to ask me to buy my needle to be tested. In India, I used to take needles with me every time I needed to be tested for anything because I know how things run there.

Then the medications used in PHCs are very few. We used to treat anything and everything with acetaminophen (Tylenol). In addition, I have never seen HIV medication in my five years of medical education prescribed to anybody.

Dr. Pace I thought HIV medications now are readily available in India and cheaper than most of the other countries.

Dr. MM These medications are available in places ready to

deal with HIV, like some hospitals or organizations, which deal with HIV patients. Many of these places are honest efforts to control this disease but in places for general population these medications are not available and, if available, patient has to buy them and many patients can't afford.

Dr. Pace Did anybody talk about it in news?

Dr. MM Everybody knows the problem is not in the PHC level only. It goes very high up to the level of the minister himself. If the government allots 100 million for supplies to PHCs, the minister will take some money somehow and allot, for example, only 80 million; the people under him will steal money and allot 50 million, and by the time it comes to purchasing the supplies only 10 million remained. Then, in PHC, 80% is sold in black market, and only 2 million in reality is given to people. So, if I ask why I have less needles or supplies, then they will say we got less needles from the government, and if I ask the lower people in government why I got less money, they say ask the higher people, and keeps going up until everybody is involved.

Dr. Bagasra Corruption everywhere, everybody involved.

Dr. MM I remember somewhere, in early 2000, somebody said that HIV India somewhere in 2020 will be four percent or some number; I am not sure and all the newspapers started criticizing him–how do you give this number–this will mean millions and millions of Indians will have HIV.

Dr. Bagasra Tell us now about Muslim discrimination in your

college and how Indian government tries to show Muslims in less number than what really is.

Dr. MM Indian government says Muslims are around 12-17%, as I remember, which many Muslims believe is far less that the truth. The reason is that they want Muslims [to] feel that they are minority, and should not try to fight for their rights. Many Muslims believe the Muslims are more than 26% in number. The percentage varies...and some states have lower numbers than others.

Then in caste system, which is the social system in India, some castes are considered high castes, some considered lower castes. The lower castes get many privileges by the Indian government to help them to improve the socio-economical status. For example, a lower caste student needs to score far less to get into medical school than a high caste student. By this, lower caste gets the chance of improving their status.

Dr. Pace So there is an advantage of being from a lower cast.

Dr. MM Yes. The strange thing, they put 80% of Muslims, even the poor ones, in middle class castes and high castes so they lose all these privileges.

Dr. Bagasra So they do not give the Muslim minority their rights and by that stop them from coming up.

Dr. MM What they say, Muslims were ruling India for long time so they cannot be in low caste. That [was] 300 hundred years ago,...now many are very poor.

Only 10% of Muslims, I guess, are considered from low caste, the Julaha are one example.

Dr. Pace — We call it affirmative action here in the US.

Dr. MM — Gandhi wanted to help the lower caste people, and made these rules to help them improve themselves, but these rules were supposed to be only for limited time, I guess 50 years. However, the government is still extending these rules because many people in the government are from low caste. It is very normal situation to see a multimillion-business man with many companies, but he is from low caste, and he and his kids are still benefiting from the laws, and his kids score less and get in medical schools and other stuff.

Dr. Pace — That is terrible!

Dr. MM — So Muslims are not benefiting from this system.

Dr. Bagasra — Tell me about your school.

Dr. MM — My medical school was established in early 1980 by some Muslim lawyer as a medical school for Muslim minority students. It was in an area heavily populated with Muslims and not much of education. Then this Hindu politician asks this lawyer to make this school for Muslim and Hindu minority because he is from a low caste and wanted to help his community. So this school then was established as school for minority in general. This politician later on kills this Muslim lawyer, with the aid of the chief of security in the medical school and the Chief of forensic department. The chief of security killed him in a train, then the forensic chief reports that he died natural death!

Later on, this school is made into school for Hindu minority and there were many riots in that area. Now Muslims are very few in that college, and many residents of that area hate the school and always want to make troubles for them. Muslims in that area lost their right as minority with no action from the government to help them.

Dr. Pace — Tell me about how circumcision can help in HIV prevention in India?

Dr. MM — [In] India the problem is the political system and everything is pulled into politics if you want the government to encourage people for circumcision you will face a lot of problem. If the ruling party agrees with you then the opposition will say no; they want us to lose our identity, and if opposition agrees then ruling party says, "no, they want us to lose our freedom." The whole idea is to get conflict to make people vote for them. Circumcision is widely practiced by Muslims, and asking Hindus to do it is, for them, like asking them to change their religion. I remember during riots in India in some places they come to know if you are Muslim or Hindu by looking if the male is circumcised or no. Similar thing happened during the civil war in Lebanon.

Dr. Pace — But in the lecture [in India, in 2006, by Dr. Bagasra, at which Dr. Pace was present], one scholar [supported] circumcision.

Dr. Bagasra Yes, and he was Hindu.

Dr. MM I agree; he will [support it] but what about the people who are the target of such idea. They are the same people on which the political system runs; they are the vote. Therefore, I think this idea is difficult. I remember one of my professors did his PhD on penile cancer, and he said that he had seen 700 cases, and after all this he did not practice it [circumcision] himself, nor his kids because of the religion factor!

Dr. Bagasra We got good response from students in our lecture.

Dr. MM I am sure you got the question on broader way to do this; it might get difficult and sometimes impossible.

Dr. Bagasra Thanks MM.

# chapter nineteen

# A Note on Needles: HIV/AIDS, India, and a Revealing Interview

(Donald Gene Pace, Krishna C. Addanki, Omar Bagasra)

AIDS-a (AIDS in Croatian)

Interview
Unsettling, unethical
Using, contaminating, reusing
Unclean needles, vicious infections
Discovery

OVER THE PAST DECADE, researchers have sounded a collective alarm about various needle-use practices in India. In 2000 [183], Singh, Dwivedi, Sood, and Wali wrote about "the re-use of needles by local medical and paramedical practitioners for administering anti-leishmanial drugs," and warned that "this trend, if not checked immediately, may have drastic consequences in the horizontal transmission of HIV in Bihar." Two years later, Sood, Midha, and Awashthi [184] spoke out against the "sizeable proportion of family physicians in the Punjab state" that practice needle and syringe reuse, and concluded that "information on the virology, clinical presentation, diagnostic tests and management approaches were lacking among a substantial proportion of family physician[s]." In a 2003 study [185] of needlestick injuries,

Wig found that "most of the needlestick injuries were neither reported nor investigated." The likelihood of such injuries obviously increases when the number of prescribed injections increases. According to a survey conducted by Murhekar, Rao, Shosal, and Sehgal in 2005 [186], an injection was part of 18.8% of all prescriptions studied, and the average number of annual injections per capita was three. Consequently, they recommended "remedial measures, such as education of prescribers to reduce the number of injections to a bare minimum." In a 2005 [201] report on a six-year study of needlestick injuries sustained by healthcare workers in Mumbai, Rodrigues, Ghag, Bavia, Shenai, and Dastur observed that 380 injuries resulted in 50 cases of testing positive for Hepatitis B, Hepatitis C, or HIV (15 cases). The following year (2006), Gupta and Boojh [182] published an alarming report on the sloppy and dangerous biomedical waste disposal issues at Balrampur Hospital, which they call "a premier healthcare establishment in Lucknow." These authors recommend that not only Balrampur, but also other healthcare facilities, improve their biomedical waste disposal procedures. In this chapter, we focus on yet another issue related to healthcare India: needle reuse.

On 12 February 2007, two Claflin University professors, Dr. Omar Bagasra and Dr. Donald Gene Pace, interviewed a Middle Eastern physician from the United Arab Emirates (UAE) who had attended medical school in India between 1998 and 2004. His interview, conducted at Claflin University in Orangeburg, South Carolina, United States of America (US), revealed numerous non-medical issues related to the medical challenges associated with HIV/AIDS in India. For purposes of anonymity, we have referred to this physician as Dr. MM. The interview is significant, not only for what it says about HIV/AIDS prevention and treatment in India but also for what it discloses about health care in that nation. The previous chapter presented the interview itself to readers; the present chapter focuses primarily on one prominent issue that was discussed in the interview, the issue of infected needles in an unlikely place: India's medical facilities.

Dr. MM is fluent in Arabic, Urdu, and English (the language in which the interview was conducted), and also understands Hindi, the official language of India. This knowledge helped him to understand the speech and nuances of meaning of those with whom he worked as he pursued his medical education and training in India. During his years in medical school in that nation, Dr. MM served in both urban and rural areas and saw, firsthand, problems related to India's responses to HIV/AIDS in both governmental and private medical institutions. His personal opinion is that the prevalence of HIV in India has been underestimated and that there has been an unfortunate relationship between the government's concern with population issues and its reluctance to give HIV appropriate priority: "I guess Indian government has a strange policy to ignore any crisis where it can shed some population, and that is why in most of the crises their response is always late." He is concerned that the government is "not taking the right measures, so people will die and only the fittest will survive." Such an attitude reflects the type of unfeeling neglect advocated by the United States philosopher William Graham Sumner. This social Darwinist opposed what he regarded as unnatural encouragement for "the bad ones" to persist, to live. Somehow claiming divine support for his logic, Sumner declared, "The law of the survival of the fittest was not made by man and cannot be abrogated by man. We can only, by interfering with it, produce the survival of the unfittest" [187]. This sort of unfeeling and irrational thinking is not uncommon, and it is similar to the perplexing position of the United States government during the Reagan presidency (1981-1989), that refused to acknowledge the existence of an AIDS epidemic in the US [188]. Even among the general public, elitist and ignorant attitudes proclaimed, "it is a gay disease, let the sinners die, they brought it upon themselves." Unfortunately, similar attitudes have echoed across India [189].

Dr. MM notes that rural areas do not always receive "enough medical supplies from the government, but this is not the major problem." He argues that the main challenge is avertable: "most of the supplies to these PHCs do not make it to there and are sold in black market!" He laments that the "PHCs get needles and drugs and other stuff as supplies from government" and maintains that these supplies should be available, for free, to those living in rural villages. However, such is not the case. He recounts "working in the PHC in some remote area where we used to have two needles in dirty boiling water and these needles were used on 50 to 100 patients a day until it broke; then new needle is opened."

The scenario in India is frustrating not only because it involves corruption *per se*, but because it is a trust-betraying public infection that oozes with contagious consequences. It is neither blind nor accidental, but a symptom of deeper ills. Ironically, it is carried out within the very public health institutions that the public has a right to trust. The dubious needle-sanitation practices described by Dr. MM are all the more scandalous given the known and widely publicized successes of needle exchange programs (NEPs) in various parts of the world, including India itself. In this South Asian nation, as elsewhere, the tension between punishing drug abusers and practicing harm reduction (HR) reflects an ongoing response to a complex issue [190]. Yet harm reduction has been proven to dramatically reduce the incidence of HIV infection. During the five year period 1988-1992, Tacoma, Washington (US) kept its HIV infection rate for injecting drug users (IDUs) below 5%. Meanwhile, with only limited needle exchange programs in place, New York City saw its infection rate among injecting drug users soar from a serious 10% to a catastrophic 50%. Currently, some 73% of Tacoma's injecting drug users acknowledge that they have altered their drug injecting behavior to prevent contraction of HIV [191]. In India, needle exchange is difficult in the countryside because of distances and because of the fear injecting drug users have of law enforcement personnel [190]. Clean needle use in medical facilities, so theoretically simple, has proven to be a life-threatening challenge. Harm reduction is most effective, Rajkumar reminds us, when it is most complete: "Ex-IDUs as role models play a real important role in leading the current users to total abstinence, which is indeed the ultimate goal of Harm Reduction." Unfortunately, there is truth to Dr. MM's complaint: "if you ask for new needle they will tell you there is none!"

But are not these medical professionals? Dr. MM replies, "not usually, most of the doctors live in towns and cities, go to PHCs by car or bus in morning, and come back by evening." He adds that the "person in charge" is typically unprepared for such serious responsibility, "some tenth grade graduate who takes care of the building and the store" and "is from the same village and lives near the PHC." Relatively limited education, however, is not a good excuse for the avaricious use of materials provided to for the protection of individual and public health. Trampling with one foot on moral responsibility and with the other on common sense, such an individual "sells all the free needles and drugs in black market to make money and sometimes some doctors take share in it. If the needles are not sold in black market," they are used to make additional profit at patient expense: "sold in the same PHC to the patients, in case a patient demands clean needle they instruct him to buy it from them!"

These supplies come, in part from the United Nations (UN) and other agencies; tragically, much of the goodwill is squandered when vital supplies never reach their intended audience, the people of India. By raising the issue of clean needle availability and usage, Dr. MM has raised a critical issue. Whether within Primary Health Centres or on the streets, new needles prevent disease, in the patient or the injecting drug user. As Stine reminds us, "NEPs are preventing HIV infections in drug users, their sex partners, and their children" [191]. Critics would do well to at least consider that behind the visible arm of the IDU are invisible partners and children; a single clean needle can protect more than the IDU into whom it is injected. Injection need not mean infection. A 12 January 2008 report from New Delhi explained that the World Bank (WB) has also expressed serious concerns about corruption in the use of the extensive funds it provides for India. The title of an article in *The Economic Times* says much: "WB upset at corruption taking toll on its health projects." In an official statement, the World Bank stated, "A detailed implementation review (DIR) launched by the World Bank in 2006 and supported by the government of India found serious incidents of fraud and corruption in five health projects. The government has announced its intention to re-examine ongoing and future projects to ensure they incorporate the lessons from the DIR." This detailed implementation review reflected concerns over major funding for malaria control ($114 million), a health systems project ($82 million), a food and drug project ($54 million), tuberculosis control ($125 million), and control of HIV/AIDS ($194 million) [192].

The practice of boiling needles to ensure cleanliness is obviously better than taking no precautions at all, but even this procedure is not always followed, and quality control procedures are sadly deficient [193-194]. Dr. MM explains that "they claim the water is boiling so the needles are clean but the problem is when the water is not boiling, which happens very often." This is not necessarily a function of human carelessness or malice because the boilers used to clean needles are powered by either "gas or electricity and in either case if gas is over or electricity down," a frequent occurrence, "the water is not boiled and needles are still used" in spite of the risk such negligence imposes on individual patients directly, and on the larger community indirectly. Unfortunately, even if power is available to run the boiler and an attempt is made to cleanse a needle, this still may prove insufficient if medical personnel, for any reason (including fatigue or high patient demand), withdraw the needle too soon from the water, or if the water is not hot enough for adequate sterilization. The heat involved in the procedure may produce an appearance of sanitary practices, but routine cleanliness, not appearance, must be the norm.

If one seeks to know the whereabouts of the government-supplied needles, an evasive response, like this one, may be given: "the government did not give any needles this month or the last three months," although later the truth emerges that the needles have already been "sold in black market." When asked about the purchasers of these needles, Dr. MM said they go to "drug shops outside the village and sometimes to the patients attending the PHC if they want clean needles or drugs." In other words, if illegal drug injectors or legal hospital patients want clean needles, they may very well need to buy them personally, perhaps on the black market. Dr. MM's comment is all too true because in India, unlike the US, one can purchase drugs from a pharmacy with no prescription. Consequently, those in the upper or middle classes, who are aware of HIV risks, can generally purchase their own drugs and clean needles. However, this leaves out India's majority, it abundant poor, who tend to be less informed about health risks, lack money to acquire quality healthcare, and are at the mercy of doctors that they trust are caring for them. Moreover, as Dr. MM added, "Some of these persons in charge of PHC run their own drug stores; they have enough drugs and supplies to run a drug store!" Knowingly suspicious of the irregular needle sanitation practices, Dr. MM was careful to watch out for his own safety: "I remember as a foreigner I was instructed to have an HIV test within 40 days of entering India and whenever I used to go to the center of testing, they used to ask me to buy my needle to be tested. In India, I used to take needles with me every time I needed to be tested for anything because I know how things run there."

In a 2005 report to the United Nations, India's National AIDS Control Organisation (NACO) issued an informative report in a special session that focused on the HIV/AIDS epidemic. The 62-page report mentions needles 14 times as it speaks of the mutually-reinforcing relationship between injecting drug users (IDUs), unclean needles, and HIV/AIDS. This timely and well-written document gives an overview of India's struggle with the epidemic, and serves as a clarion call to future action. What the report does not mention, however, is the needle sanitation scandal that festers in places where model hygiene should prevail, in India's primary health centres. As the United Nations report explains, IDUs have spread HIV to a bridge population, which in turn has transmitted the plague to the general population of India [195-198]. This lamentable pattern must continue to be considered, challenged, and checked. As the nations of the world, including India, struggle to stop the AIDS deluge, they must prioritize and publicize their efforts in a way that does not ignore less obvious streams of infection, including those that have leaked into India's primary health centres [198-200].

Because social attitudes commonly influence health and medical practices, we urge governmental policymakers, health officials, and private citizens to remember that the transmission of HIV and the development of AIDS are preventable and ultimately spring from unwise human choices. Consequently, public policy must focus on *both* biological and behavioral aspects of the modern-day epidemic that directly threatens one-sixth of planet Earth's inhabitants (i.e., the nation of India), and indirectly menaces the rest. National political boundaries do not double as barriers to disease. India's problem is also humanity's concern. HIV/AIDS never carries a passport, has no visa to stamp, and yet crosses national boundaries at will, as if to mock imprudent human behavior, and prove that the planet's geomedical dilemmas are at least as complex as its geopolitical challenges.

Acknowledgments: This study was supported by grants from NIH/NIMHD, 1P20MD001; NIH/NICRR, 2P2RR16461; NSF, EPS-044760; and UNCFSP-special programs (OB).

# chapter twenty

## Confession and Complaint: Bad, Worse, and Worst

এইডস রোগ (AIDS in Bengali)

| Faith |
|---|
| Courageous, compassionate |
| Listening, teaching, loving |
| Individual effort does matter |
| Works |

THIS CHAPTER IS BASED substantially on an interview with a physician (referred to anonymously as Dr. H.) who was instrumental in persuading sex workers in Hyderabad, Pakistan to transform their lives and educate others about the devastation immorality wreaks on both the human body and the human character. The interview, summarized by Dr. Bagasra and later edited by Dr. Pace, took place on 13 April 2011 at the National AIDS Control Center in Karachi, Pakistan after an ad hoc meeting of the NACC earlier that day.

In her perceptive little volume about confession and literature, Spanish author María Zambrano argues that crises provoke confessions. Confessions are evidence that humans are trying to find themselves, particularly when weighed down with humiliating burdens, or by personal failures. Confession doubles as a means of coping and escape. The despair that inspires confession may spring from guilt, but it is also, like the anguished confession of the Old Testament sufferer Job, "queja, simple queja" (complaint, simple complaint) [202]. The complaint of a confessant is accompanied by hope, a belief that wholeness may yet be possible. Confession is a literary, and living, genre of angst and hopefulness, of death of the old life and birth of the new [203]. The confession-complaint of this sex worker from Hyderabad, Pakistan is both poetic and pathetic: "I am like a fire hydrant where every dog comes to relieve himself." Dr. H's rapport with these workers allowed her to talk with them, and perhaps even more important, to listen to them, to be their confessor as it were. Her care for these women allowed her to become an understanding teacher to them, one whose goal was to lift them to a higher and safer plane. Her concern to protect these women from self-serving "dogs" seems to have brought greater understanding and safety. Confession and change are common in every nation and every culture; they have a place in the quest to prevent HIV infection. Change is possible, and desirable, even if it comes in small packets. While communities must take measures to preserve the innocence of the innocent, they must also actively urge change; they must urgently promote the dismantling of risky behaviors and hazardous institutions.

Dr. H., a beautiful and remarkable young Pakistani physician from a conservative Muslim home, became a sex worker's confidant and educator. Her desire to curb the increasing devastation inflicted by HIV/AIDS on this chaotic Islamic nation superseded all her reservations and fears. Intent on stopping the downward slide she determined to do something about her dream. "I came from highly educated Syed home [i.e., a family that claims ancestry back to the prophet Muhammad]. My father was minister of education for many years in the Sindh Providence. Two of [my] sisters are also physicians; one resides in Toronto, Canada. After medical school, I wanted to work as a physician, and somehow I decided to work in the red district." In the interview with Dr. H., this young physician recounted the genesis of her desire to make a difference in her society. She confided in her supportive father, also a determined proponent of reform, who helped her locate a small room, and set up a clinic, in the heart of the red district of Hyderabad. She recalled how she initially decided to work in an area where Hindu leather workers had settled. These middle-class craftsmen, traditionally proud of their profession, were located adjacent to sex workers, called *Khothas* (a term that denotes a small mansion but in practice is generally used as a word for brothels).

Dr. H. has rather comprehensive insights into the sex business in Hyderabad, the second largest city in the Sindh province of Pakistan, and the fourth largest city in the country. The city is home to about two million people from all walks of life; Sindhi (the native folk), Pathans, Muhajirs (immigrants from former British India), and Punjabis live in the city. The majority of the sex workers in the Sindh province belong to this latter group. The city of Hyderabad was founded in 1768, far earlier than the creation of Pakistan in 1947. Known prior to the 1947 Pakistan-India partition as the "Paris of India," the city is noted being perhaps the most colorful city of the entire nation, a city rich in industries and craft production. The colorful dresses of the women, regardless of wealth or status, have become a distinguishing characteristic of this urban center.

In this fascinating Pakistani city, Dr. H. made the neighbors aware that she was a doctor, and that she could treat everyone for free, something she could do because of funding received from a government grant designed to help reduce sexually transmitted diseases, including HIVAIDS, particularly in sex workers. Initially, she predicted that no one would come. For days, only rarely did anyone enter her clinic, but even these were not the people she was trying to reach. Then slowly, slowly, little children, mostly boys of a tender age, could come and chat with her; these were the children of the red district. Afterward, some older ladies, madams, started to come. They inquired about her motives: "Why would you treat us for free?" They questioned whether she was endeavoring to extract HIV-related information from them in order to expose them to the media, particularly the newspapers? They had been bitten by this kind of snake previously, and they were wary of future public exposures. Men had come before to draw blood, to test for disease, and then they announced that HIV-infected workers were among their numbers. There was the shock of the AIDS announcement, and troubling financial repercussions to their business. Word of the infections spread quickly; stigmas, already plentiful in red-district business, became even more negative.

The physician, with a guileless confidence born of pure motives, explained to them that she was not interested in pursuing the paths that had troubled them earlier. She was a doctor that just wanted to help. Early on, she avoided the mention of AIDS, even indirectly. She began treating their children for their fevers; she began assisting mothers with their deliveries. She knew that these unsophisticated common people did not tend to go to doctors but to midwives, and that these women listened to them, heard their confessions as it were. She kept their secrets; she listened to their woman-to-woman complaints. Midwifery runs in families; it is multigenerational, with all the benefits to society of a cherished and trusted institution. Unless a midwife recommends that a pregnancy is high risk, none of the sex workers would ever go to hospital or a private OB/GYN office.

In the beginning, Dr. H. would ask a mother for the name of the father of their child. That was a mistake, and she realized it quickly. It was a bad idea to even ask. These children have no fathers in their life. If they have to be hospitalized for some reason, they would use an uncle's name or the name of a male who is closely related to them. Dr. H. slowly gained their trust. She helped them during their times of grief and sorrows; she participated in their rituals; she assisted with difficult deliveries; she consoled them after miscarriages. She got to know them; she understood their way of life. "These folks are wonderful people," she commented. "They practice religion, they read [the] Qur'an, observe five-times-a-day prayers, and generally follow [the] Shi'a school of Muslim law. During the last ten days of Maharam [a holy month for the Shi'a branch of Islam], they close their business and reflect and mourn the death of Imam Hussain [ibn Ali, the Prophet Muhammad's grandson who was assassinated by the Umayyad king]. They accept their destiny as sex workers, and do not have remorse." Subsequent actions, when given better options for their lives, revealed that the women did have some deeply-experienced remorse, and were willing to challenge their apparent destiny.

Dr. H. related how the madam initially wanted to know if she wished to close their business, and take their workers away from them, but she was able to convince the madam that she contemplated nothing of the kind. Instead, she began to teach them about sexually transmitted infections (STI), the symptoms of the infections, and how to detect them. Eventually, she started a full-fledged, government supported HIV/AIDS and STI prevention clinic in the red district. Dr. H. hired women from the same community, a peer group approach, whom she would train. They would set up education camps, but do so secretly for enhanced impact. Only ten women were allowed in each group, and the groups were separated into different rooms. Here they were tested for their ability to detect symptoms of the various STI. Here they were also rewarded, with gifts of their own choice, when they responded with correct diagnoses. They could choose such items as lipstick, talcum powder, or a small make-up kit. The girls were excited! They always wanted to learn new things, longed to expand their horizons. Weary of abuse to their bodies, they also welcomed the opportunity to use their minds.

At the beginning when she was still moving along cautiously, Dr. H. refrained from using the word "condom." At that time, the use of such a protective device was considered taboo in their business dealings. She knew, and wanted them to know, that such protection could be a matter of life and death for them. These workers had been indoctrinated instead to have unprotected sex with their clients, and to never question their perilous practices. The young doctor made them aware of the enormous protective benefits of condoms, and did so gradually through patient and friendly persuasion. It was the older generation that resisted change; the younger workers came aboard quickly.

Still, the world of the red district was one of tragic choices, selections that were not between good, better, and best, but between bad, worse, and worst. Dr. H. tried to help them make life-saving decisions in their live-threatening occupation. Still, when they found that many of their "dogs" did not want condoms, they resorted to clever methods to get them to use them anyway. Although it is not universal, a high percentage of sex workers and clients use alcohol [204], so the workers learned to give more alcohol to the clients than what they would drink themselves in order to make them more amenable to condom use. After that, they could use either female condoms or easily slip one on onto the male client. In this world of bad, worse, and worst decisions, some of the workers have desired to become pregnant themselves, and have selected a handsome, fair-skinned, and robust man to father their child. Unlike the overall practice in South Asia and China, they have often preferred female babies over male ones, since they believe females will brings security to their future. They stereotypically consider males to be docile and lame, individuals who irresponsibly depend on the financial support of their females.

Of course, despite Dr. H's personnel success, and the satisfaction she feels from knowing she is helping an often helpless and tragic community, she has to pay a personal price in terms of her own reputation. Many times, family or friends have seen her coming in and out of the red district, which prompted them to inquire why a decent person like her, a lady from a conservative Muslim home, would work with such sinners. She has answered them that these unfortunate and abused workers are also God's creations. She has asked them what they might have ended up doing if they had been raised in similar circumstances to those whom they now so quickly condemn. They dislike her responses, continue to judge her harshly, and remain unconvinced and unsympathetic to what Dr. H. is trying to do.

The doctor explained that the sex workers in the city can be divided into three major categories: those who work in the brothels, the street hustlers, and the Kotikhana. The brothels are confined to the red district of the city. The sex workers there are commonly Punjabis, and are professionals who form a somewhat caste-like community. The workers in this group are often part of a multi-generational group; their work is a type of unfortunate inheritance. Every female in the group works for the same goal, with older women managing the younger generation. Far from the practices of parents and teachers in most settings, who urge upright behavior, and warn against evil books and media, these women tutor the next generation in the opposite direction, and thus provide them with skills that will unfortunately keep them trapped and miserable, and even lead to their premature deaths. Rather than ignore these individuals, Dr. H. has sought to at least try to improve their plight, and teach them how to protect themselves from the exploitation that characterizes their trade. Protection against HIV in this group also helps to protect the larger society.

Times have changed and techniques have become both more sinister and more sophisticated. Girls often go to designated private places that are preferred by their clients. The street hustlers used to look for business at their own specific corners of the city blocks, but their activities have evolved into a network. These are a totally independent group of sex workers not related to those described above. They are generally educated girls, college or health care industry girls. Yet in spite of the warnings their education and moral instruction should provide, they promote practices that endanger not only personal and community virtue, but also the health and moral climate of their city and surrounding areas. They can be very sophisticated with their iPhones, and their internet access. The middlemen they rely on can also be sophisticated in their methods. Although intelligent exploiters of females for personal pecuniary gain, they seek to minimize the level of exploitation of the very women they place in abusive situations. They may even be well-armed, in order to protect their nefarious investment. This group generally avoids any kind of HIV test, or other precautions, that may save their own lives or those of their clients.

In the world of bad, better, and worse we are describing, the third category seems to be the worst. This home-based business is occult, and even more evil than its foul counterparts. A madam operates her exploitative business, for example, by purchasing four to five young girls from Punjab. These become her slaves, her personnel property. A madam would pay for these girls according to how she perceives their outward beauty. With demonic skill, she then rents a nice house in a prosperous neighborhood, and sets up the business. The neighborhood is never the same while her business is there, but usually she runs a short-term rental operation, and then moves quickly to other areas if it becomes obvious to the neighbors what she is perpetrating in their neighborhood. Sometimes one of the girls may run away but not for very long. These people are highly connected and will reclaim their property quickly with severe consequences for the girls who seek to escape from their bondage. A headhunter will bring the girl in with satanic zeal. A variation of this home-based approach also mocks the traditional images of a home as a place of love, refuge, and safety. In this variation on a tragic theme, a group of sophisticated, educated, and upper-class girls charge as much as 10,000 rupees per night. Like perverted models, they mask their iniquity with expressive designer clothing, and luxurious cars. These are girls – one would not call them ladies – who sell themselves for money and personal enjoyment. They certainly do not need the money. These delight in sinning with other people of high status, including politicians, sports stars, and other wealthy individuals. Drugs, including alcohol, are often part of their world.

In all variations of this devious business, payoffs to law enforcement play an important role. Islam does not promote immorality, and given the predominant place of Islam in Pakistan, such business should, technically, not even exist. This routine "housekeeping expense" threatens homes and health in Pakistan and elsewhere in the South Asian subcontinent. This business, in whatever form, defies the spirit of Islam, Hinduism, Sikhism, and Christianity. It takes a certain type of courage for a Dr. H. to look beyond the corruption, and the moral stench, of the sex business, and seek to protect the lives of individuals who are in virtual bondage. Yet Dr. H. has sought to move those she cares about, and seeks to serve directly, from worst to worse, or even from worse to bad. It is a pathetic world but Dr. H. sees some reform as better than none, and hopes to do at least something to counter the spread of HIV/AIDS in a largely clandestine environment that encourages immorality, drugs, alcohol, and irresponsibility, the very practices that threaten the virtue and the very lives of private individuals and the body politic.

# chapter twenty-one

# THE LANGUAGE OF AIDS

AIDES (AIDS in French)

| Words<br>Zulu, Xhosa<br>Recognizing, fearing, conveying<br>Expressions spring from experience<br>Meaning |
|---|

WHAT AFFECTS SOCIETY AFFECTS language, and the global concern over AIDS is no exception. A look at how some African languages describe AIDS gives insights into social impacts of this epidemic. Metaphors say something about human creativity, and patterns of expression, but they also reflect how individuals envision HIV/AIDS. Professor T. Dowling's study of the Cape Town, South Africa area is particularly illuminating in this regard [169]. Because of its geographic focus, Zulu and Xhosa expressions predominate. In Africa, giving names is routinely done so with meaning in mind. The name *Ntombizodwa*, for instance, means "girls only," and applies to a child in a family with no male children. A child that comes after a long period of waiting may be named *Lindiwe*, the word for "expected" [169].

Names that praise an individual or thing are also used, as seen in the Xhosa poet Bongani Sitole's tribute to Nelson Mandela.

Hail Dalibhunga!
Words of truth have been exposed,
He's the bull that kicks up dust and stones and breaks antheaps,
He's the wild animal that stares at the sky
Until the stars fall down [169].

The names attached to HIV/AIDS bear similarities to the naming patterns used for leading personalities and mighty leaders.

Compare the praises of a great fighter, Mqikela Ndayi, a notable fighter, is called by a military metaphor: "He is a rifle speeding to its target." HIV/AIDS, a notable killer, has been given a similar name: "He who shoots to kill" [169]. A person living with AIDS is one who has been suffered: "a hot coal fall upon himself or herself." Such a person has "caught it" [169]. AIDS is "the One who Finishes the Nation," or "The Finisher of the Nation." It is "The Killer of the Nation." The names of AIDS are similar to those assigned to prominent Africans. The word for AIDS in Zulu and Xhosa is similar to that for the word plague. South Africa has eleven official languages. In English AIDS is "a disease for which there is no cure"; the word kill is not part of the description. The same is true in Afrikaans. Such is not the case with the nine other languages. Here, the Xhosa description is typical: "a disease which kills and which cannot be cured" [169]. Slaughtering of animals commonly accompanies funeral rites, and HIV/AIDS has been described as an illness that plagues the dwelling place of the living, "contaminating" a place such as KwaZulu-Natal, now styled KwaNyama-ayipheli, "at the place where the meat does not end." The reference to meat is not a symbol of prosperity but of the demand for meat at funerals. "Sodom and Gomorrah, a place from where you will not return" is another death-associated name for KwaZulu-Natal. *Kwatsi*, which translates as "disease" in the Sotho language, is linked in its provenance to a deadly disease that can spread from cattle to humans: anthrax. *Bacillus anthracis*, the scientific name for the anthrax organism, forms spores and can persist in very challenging conditions. It is highly resistant to cold and heath, to drying, and even to chemical disinfectants. Anthrax spores can live on and on in the soil [169, 18]. The Sotho phrase "he or she is being held by HIV/AIDS" uses the word "kwatsi," the anthrax-related word for disease. To be held by AIDS is, with gloomy symbolism, to be held by *kwatsi*, by anthrax [169].

In African language, AIDS is personified as a killer: "The Beater-up of People," "The One Who Shoots to Kill," "The Finisher of the Nation," "The Indiscriminate Killer," "The One Who Chops down." With a kind of gallows humor, other linguistic expression portrays it as a type of game, or as a train ride on an already overcrowded train. Zola and uMlazi are African place names, the former located in the Soweto area, and the latter in KwaZulu-Natal. Thus, when it is said that the Zola or uMlazi railroad lines are overcrowded, the upshot is that these populous areas have high rates of HIV infection. Sometimes, the words for HIV or AIDS are simply avoided, as with a taboo. HIV/AIDS has been called, in reference to its being an acronym, "four words," or "three words." In reference to its seriousness, HIV/AIDS is "big words" or "the big matter." The relationship between actual diagnosis and anxiety is also reflected in African naming patterns. Who is it that is afflicted with AIDS? A common response in the Zulu language runs contrary to accurate cause-effect logic but is, nonetheless, understandable: "People who go for tests" [169]. Like a person who is cheerful and hopeful until they find out from a doctor that they have cancer, so HIV-infected persons often have live without worrying. Following diagnosis that they are HIV seropositive, however, they then have both the infection and the worries that go with it in its diagnosed state.

Some Africans have adopted a conspiracy mentality with regards to AIDS. They accuse foreign nations of spreading AIDS to Africa: "Our leaders who travel to all these countries come with Aids." Race relations also influence the accusations:

"Whites came to Africa with food that has Aids in it"; "The different nationalities that are filling up SA are those that are coming with these diseases" [169]. Many have also perceived a sinister foreign plot to end love among Africans, an attitude that implies racial mistrust as well.

Lottery imagery is used to refer to HIV/AIDS. The chances of winning the lottery are very small; those infected with HIV also have little hope of winning. "She is playing the lotto" is one allusion to an infected person; she or he "has the lotto" is another. Red-scarf imagery is used in reference to the symbolic AIDS ribbon or a red scarf: "She is wearing something round the neck" or "She is wearing the red scarf." AIDS has also been called "the disease of beautiful people." It brings great loneliness, and once who lives with AIDS is a "cow who eats alone." That AIDS routinely brings life to a premature end is tragic enough, but this ultimate consequence is made sadder along the way by the effects of AIDS on personal relationships: "It ends love" [169].

# chapter twenty-two

# THESE DIED OF AIDS

HIESaren (AIDS in Basque)

Exempt?
Prince, pauper
Living, lamenting, shriveling
Death imposes morbid equality
No!

BERRY BERENSON DIED ON September 11 (the September 11) aboard American Airlines Flight 11, the victim of sinister terrorist plotting. Her husband, Anthony Perkins (1932-1992), died about a decade earlier of AIDS-related pneumonia. Perkins was closely associated with actor Rock Hudson (1925-1985), who acted in almost 70 movies before dying of an AIDS-related infirmity [170].

Keith Haring (1958-1990) painted a huge mural on the Berlin Wall at an internationally famous landmark, the Brandenburg Gate before he died of AIDS-related causes in 1990, and before a number of his acquaintances also died of complications related to AIDS. [170]. Africa is home to nearly 15% of the population of the planet, and in 2009, nearly three-fourths (72%) of all AIDS-related deaths occurred there [170].

"Beauty and the Beast" the movie featured an Academy Award winning song by the same name. Howard Ashman (1950-

1991) wrote the lyrics before he died of AIDS-related causes. The Academy Award for the song was given to him in 1992, posthumously [170].

A child prodigy, the flamboyant piano genius Liberace (1919-1987) was the world's highest-paid performer for a lengthy period of time. He first lost his health (which AIDS attacked), then his wealth (which death took from him) [170].

Born with an incredibly versatile voice with a vast range, Farrokh Bulsara, or Freddie Mercury (1946-1991) lived only 45 years; it was difficult to live longer with AIDS, a disease he disclosed the day before he died [171].

After his eldest son, Makgatho Mandela (1951-2005), died of AIDS-related causes, AIDS activist and former South African president Nelson Rolihlahla Mandela (1918-) continued to lobby for AIDS awareness and funding. Mangosuthu Buthelezi (1928-), another noted South African political leader, lost two of his children to AIDS-related illnesses prior to the passing of Makgatho Mandela [173-175].

One of the most prolific authors imaginable, Isaac Asimov (1920-1992) wrote over 500 books before he died of AIDS-related causes. He might have written more had he lived free of HIV [176].

French interdisciplinary philosopher Michel Foucault (1926-1984) wrote of power and health policy, introduced new ways of thinking, and taught at the University of California at Berkeley before he died at age 57 of AIDS-related causes [19].

Hemophilia, a serious bleeding disorder, challenged Ryan White (1971-1990), but when a blood transfusion gave him HIV, his challenges became much more severe. Before dying of AIDS-related causes at age 18, Ryan challenged school authorities about his right to attend school, and helped educate numerous people about AIDS myths, and taught them about courage along the way [20].

After dying from AIDS-worsened pneumonia, his body lay in state at the Richmond, Virginia Governor's Mansion, an honor not given to anyone since Civil War General Thomas J. "Stonewall" Jackson's body lay there. Arthur Ashe (1943-1993) was a fabulous tennis player, and a caring human being who helped shatter AIDS stereotypes when he died [172].

AIDS is a human disease, and it does have a human face.

# chapter twenty-three

## Etcetera: These are the Houses that AIDS Built

艾滋病 (AIDS in Chinese traditional)

| Casket<br>Wooden, simple<br>Chopping, hammering, burying<br>Simple design, hand crafted<br>Residence |
|---|

THIS IS THE COFFIN that gave the husband who succumbed to AIDS a new-yet-smaller wooden residence.

This is the tree that supplied the wood to build for the husband who succumbed to AIDS a new-yet-smaller wooden residence.

This is the son who chopped down the tree that supplied the wood to build for the husband who succumbed to AIDS a new-yet-smaller wooden residence.

This is the HIV-infected, grieving spouse who gave birth to the son who chopped down the tree that supplied the wood to build for the husband who succumbed to AIDS a new-yet-smaller wooden residence.

This is the tree-chopping orphan of the now deceased HIV-infected, grieving spouse who gave birth to the son who chopped down the tree that supplied the wood to build for the husband who succumbed to AIDS a new-yet-smaller residence.

Etcetera.

P.S. This is HAART, but there is never enough, and it only delays, never cures.

# chapter twenty-four

# AIDS Personified

ఎయిడ్స్ వ్యాధి (AIDS in Telugu)

**SIDA**
**Għajnuniet,** 에이즈
**UKIMWI,** الإيدز, СПИД
СНІД, ایدز, СПИН, ఎయిడ్స వ్యాధి
**AIDS**

AIDS IGNORES POLITICAL BOUNDARIES, crosses porous borders of poor human choice, pays no customs duties, travels with neither passport nor visa, drives the world's highways, sails the planet's seas, flies through earth's airspace, breaks hearts, breaks down immunity, covers truth, spreads myths, promotes ignorance, wrecks homes, grieves children, fathers orphans, leaves widows, chops trees for coffins, exploits poor choices, gnaws away at health, robs the rich, impoverishes the poor, destroys CD4+ counts, baffles scientists, challenges budgets, reduces profits, encourages absenteeism, promotes burials, worships immorality, upholds double standards, exploits women, exploits men, exploits children, flourishes in an apathetic NIMBY (not in my backyard) environment, avoids unified NIOBY (not in our backyard) efforts [177], evades classical immunity, avoids slumber, takes no vacations, torments the adults in Cape Town and Mumbai, confuses the children in orphanages and on streets, dries no tears, sheds no tears, distributes no handkerchiefs, makes no mortgage payments, slays mortgage payers, robs down payments, kills down payers, evicts tenants, plays Trojan horse, shares freely, terminates employment, drives up insurance rates, drives down population growth, loves drug users, mixes well with alcohol, encourages fear, squelches hope, creates morbid personal histories, changes the history of nations, tortures the present, and slays the future.

# NOTES

ՁԻԱՀ - ի (AIDS in Armenian)


1. Paine, Tom. *Common Sense.* Great Books Online. http://www.bartleby.com/133/ <accessed 28 November 2011>.

2. Stowe, Harriet Beecher. Uncle Tom's Cabin. The Literature Network. http://www.online-literature.com/stowe/uncletom/ <accessed 28 November 2011>.

3. King, Martin Luther, Jr. "I Have a Dream." American Rhetoric: Top 100 Speeches. http://www.americanrhetoric.com/speeches/mlkihaveadream.htm <accessed 28 November 2011>.

4. Koppel, Ted. "Koppel on Television and Morality"; http://www.mediaresearch.org/mediawatch/1989/watch19890401.asp#analysis <accessed 28 November 2011>.

5. World Scripture. Adultery. http://www.unification.net/ws/theme059.htm <accessed 28 November 2011>.

6. Hindu Books Universe. Manusmriti: The Laws of Manu. 352. http://www.hindubooks.org/scriptures/manusmriti/ch8/ch8_351_360.htm <accessed 28 November 2011>.

7. Bible, King James Version. http://quod.lib.umich.edu/k/kjv/browse.html <accessed 28 November 2011>.

8. Craig C. Freudenrich, "Mosquito Bites, Diseases and

Protection," http://animals.howstuffworks.com/insects/mosquito3.htm <accessed 26 May 2011>.

9. Centers for Disease Control and Prevention. HIV Transmission. http://www.cdc.gov/hiv/resources/qa/transmission.htm <accessed 28 November 2011>.

10. "The Joy of What?" *Wall Street Journal* (12 December 1991, A14).

11. Lives in the Law: D. Todd Christofferson, '72, Duke Law: News and Events. http://www.law.duke.edu/news/story?id=4710&u=11, <accessed 31 October 2011>

12. Christofferson, DT. Moral Discipline. *Ensign*. November 2009. http://lds.org/ensign/2009/11/moral-discipline?lang=eng#footnote2-04211_000_036 <accessed 31 October 2011>).

13. Charles Francis Adams, The Works of John Adams, Second President of the United States, 228-29 (Books for Libraries Press, 1969).

14. Wu X, Zhou T, Zhu J, Zhang B, Georgiev I, Wang C, Chen X, Longo NS, Louder M, McKee K, O'Dell S, Perfetto S, Schmidt SD, Shi W, Wu L, Yang Y, Yang Z-Y, Yang Z, Zhang Z, Bonsignori M, Crump JA, Kapiga SH, Sam NE, Haynes BF, Simek M, Burton DR, Koff WC, Doria-Rose N, Connors M, NISC Comparative Sequencing Program, Mullikin JC, Nabel GJ, Roederer M, Shapiro L, Kwong PD, Mascola JR. Focused evolution of HIV-1 neutralizing antibodies revealed by crystal structures and deep sequencing. Science Express (online on Aug. 11, 2011).

15. NIH-Led Team Maps Route for Eliciting HIV Neutralizing

Antibodies: New Technique Can be Used Widely to Develop Vaccines, 11 August 2011. http://www.niaid.nih.gov/news/newsreleases/2011/Pages/HIVAntibodyEvolution.aspx. <accessed 12 August 2011>.

16. President Obama on the National HIV/AIDS Strategy. http://www.whitehouse.gov/photos-and-video/video/president-obama-national-hivaids-strategy#transcript. <accessed 2 November 2011>).

17. Zuckerman AJ. AIDS and insects. Br Med J (Clin Res Ed). 1986 April 26; 292(6528):1094-1095.

18. Stoltenow CL. Anthrax. 2000. http://www.ag.ndsu.edu/pubs/ansci/livestoc/a561w.htm <accessed 21 December 2011>

19. Wohlsen M. Foucault at Berkeley: A university transformed. Illuminations. http://illuminations.berkeley.edu/archives/2005/history.php?volume=3 <accessed 21 December 2011>.

20. Johnson D. "Ryan White Dies of AIDS at 18; His Struggle Helped Pierce Myths." Obituaries. New York Times. 9 April 1990. http://www.nytimes.com/1990/04/09/obituaries/ryan-white-dies-of-aids-at-18-his-struggle-helped-pierce-myths.html?pagewanted=all <accessed 21 December 2011>.

21. The Price in Blood! Casualties in the Civil War. Civil War Potpourri. http://www.civilwarhome.com/casualties.htm <accessed 16 December 2011>];

22. White M. Spanish Civil War. Secondary Wars and Atrocities of the Twentieth Century. http://necrometrics.com/20c300k.htm#Spanish <accessed 16 December 2011>.];

23. White M. First World War, Second World War. Source List and Detailed Death Tolls for the Primary Megadeaths of

the Twentieth Century. http://necrometrics.com/20c5m.htm#Second <accessed 21 December 2011>].

24. The Churchill Society. London. http://www.churchill-society-london.org.uk/EndoBegn.html <accessed 21 December 2011>.

25. Operation Rolling Thunder. History.com. http://www.history.com/topics/operation-rolling-thunder <accessed 21 December 2011>.

26. The Bombing of Hiroshima and Nagasaki. History.com. http://www.history.com/topics/bombing-of-hiroshima-and-nagasaki <accessed 21 December 2011>.

27. Francisco Franco. Spartacus Educational. http://www.spartacus.schoolnet.co.uk/2WWfranco.htm <accessed 21 December 2011>.

28. AVERT. The ABC of HIV Prevention. http://www.avert.org/abc-hiv.htm <accessed 29 November 2011>.

29. Addanki KC, Pace DG, and Bagasra O. "A Practice for All Seasons: Male Circumcision and the Prevention of HIV Transmission." Journal of Infection in Developing Countries 2 (2008): 328-34.

30. Schoen E. 2005. Circumcision. Berkeley, California: RDR Books, 2005.

31. Morris, Brian J. "Male Circumcision Guide for Doctors, Parents, Adults & Teens." http://www.circinfo.net/rates_of_circumcision.html. Accessed 5 February 2009.

32. Brown, David. "Africa Gives 'ABC' Mixed Grades: AIDS Abstinence Plan Raises Awareness but Has Small Effect on

Behavior. *Washington Post.* 15 August 2006: A04. http://www.washingtonpost.com/wp-dyn/content/article/2006/08/14/AR2006081401458.html <accessed 20 January 2009>.

33. Bongaarts J, Reining P, Way P, Conant F (1989) "The relationship between male circumcision and HIV infection in African populations." AIDS 3:373-377.

34. Donoval BA, Landay AL, Moses S, Agot K, Ndinya-Achola JO, Nyagaya EA, *et al.* (2006) "HIV-1 target cells in foreskins of African men with varying histories of sexually transmitted infections." Am J Clin Pathol 125: 386–91.

35. Williams BG, Lloyd-Smith JO, Gouws E, Hankins C, Getz WM, Hargrove J, *et al.* (2006) "The potential impact of male circumcision on HIV in sub-Saharan Africa." PLoS Med 3: e262.

36. Bailey RC, Moses S, Parker CB, Agot K, Maclean I, Krieger JN, *et al.* (2007) "Male circumcision for HIV prevention in young men in Kisumu, Kenya: a randomized controlled trial." Lancet 369:643-656.

37. Alanis MC and Lucidi RS (2004) "Neonatal Circumcision: A Review of the World's Oldest and Most Controversial Operation." Obstetrical & Gynecological Survey. 59(5):379-395.

38. Drain P, Halperin D, Hughes J, Klausner J, Bailey R. (2006) "Male circumcision, religion, and infectious diseases: an ecologic analysis of 118 developing countries." BMC Infect Dis 6: 172.

39. Auvert B, Taljaard D, Lagarde E, Sobngwi-Tambekou J, Sitta R, Puren A, *et al.* (2005) "Randomized, controlled intervention

trial of male circumcision for reduction of HIV infection risk: The ANRS 1265 trial." PLoS Med. Nov;2(11):e298. Epub 2005 Oct 25. Erratum in: PLoS Med. 2006 May;3(5):e298.

40. Kebaabetswe P, Lockman S, Mogwe S, Mandevu R, Thior I, Essex M. et al. "Male circumcision: an acceptable strategy for HIV prevention in Botswana." Sexually Transmitted Infections 79 (2003): 214-219.

41. Reproductive Health Issues in Latin America http://www.stanford.edu/group/womenscourage/Repro_Latin/ekobash_Homepage_Latin.html <accessed 30 November 2011>.

42. Hirsch, Jennifer; Meneses, Sergio; Thompson, Brenda; Negroni, Mirka; Pelcastre, Blanca; Rio, Carlos. "The Inevitability of Infidelity: Sexual Reputation, Social Geographies, and Marital HIV Risk in Rural Mexico." Framing Health Matters. American Journal of Public Health. (2007). Vol 97 (6). 986-996. http://ajph.aphapublications.org/doi/full/10.2105/AJPH.2006.088492 <accessed 30 November 2011>.

43. The Hebrew Bible in English. http://www.mechon-mamre.org/e/et/et0.htm <accessed 30 November 2011>.

44. BBC Religions. Religions. Circumcision of boys. http://www.bbc.co.uk/religion/religions/islam/islamethics/malecircumcision.shtml <accessed 30 November 2011>.

45. Rizvi SAH, Naqvi SAA, Hussain M, Hasan AS. Religious circumcision: a Muslim view. BJU International (1999). 83. Suppl. 1, 13-16.

46. World Health Organization (WHO), "Global Status Report on Alcohol and Health 2011," http://www.who.int/

substance_abuse/publications/global_alcohol_report/en/ <accessed 25 February 2011>.

47. Bagasra O, Kajdacsy-Balla A, and Lischner HW. Effects of alcohol ingestion on in vitro susceptibility of peripheral blood mononuclear cells to infection with HIV and of selected T-cell functions. Alcoholism: Clinical and Experimental Research 13(5):636-643, 1989.

48. National Institute on Drug Abuse (NIDA). HIV/AIDS, Research Report Series, http://www.nida.nih.gov/ ResearchReports/ HIV/hiv.html <accessed 7 October 2011>.

49. Baum MK, Rafie C, Lai S, Sales S, Page JB, and Campa A, Alcohol Use Accelerates HIV Disease Progression. AIDS Res Hum Retroviruses. 2010 May; 26(5): 511–518.

50. Baum MK, Rafie C, Lai S, Sales S, Page B, Campa A. Crack-cocaine use accelerates HIV disease progression in a cohort of HIV-positive drug users. J Acquir Immune Defic Syndr. 2009 Jan 1;50(1):93-9.

51. National Institute on Alcohol Abuse and Alcoholism (NIAAA). Alcohol Alert, No. 15 PH 311 January 1992. http://pubs.niaaa.nih.gov/publications/aa15.htm

52. MacGregor RR. Alcohol and drugs as co-factors for AIDS. Advances in Alcohol and Substance Abuse 7(2):47-71, 1988.

53. Plant MA. Alcohol, sex and AIDS. Alcohol and Alcoholism 25(2/3):293-301, 1990;

54. Pillai R and Watson RR. Response to: 'Alcohol, Sex and AIDS.' Alcohol and Alcoholism 25(6):711-713, 1990.

55. Tennenbaum JI, Rupert RD, St. Pierre RLG, and Greenberger NJ. The effect of chronic alcohol administration on the immune responsiveness of rats. Journal of Allergy 44:272-281, 1969.

56. Jerrells TR. Marietta CA; Eckardt MJ; Majchrowicz E; and Weight FF. Effects of ethanol administration on parameters of immunocompetency in rats. Journal of Leukocyte Biology 39(5):499-501, 1986.

57. Saad AJ and Jerrells TR. Flow cytometric and immunohistochemical evaluation of ethanol-induced changes in splenic and thymic lymphoid cell populations. Alcoholism: Clinical and Experimental Research 15(5):796-803, 1991.

58. Liu YK. Effects of alcohol on granulocytes and lymphocytes. Seminars in Hematology 17:130-136, 1980.

59. McFarland W and Libre EP. Abnormal leukocyte response in alcoholism. Annals of Internal Medicine 59:865-877, 1963.

60. Gluckman SJ; Dvorak VC; and MacGregor RR. Host defenses during prolonged alcohol consumption in a controlled environment. Archives of Internal Medicine 137:1539-1543, 1977.

61. Mutchnick MG, and Lee HH. Impaired lymphocyte proliferative response to mitogen in alcoholic patients: Absence of a relation to liver disease activity. Alcoholism: Clinical and Experimental Research 12(1):155-158, 1988.

62. Guarneri JJ and Laurenzi GA. Effect of alcohol on the mobilization of alveolar macrophages. Journal of Laboratory and Clinical Medicine 72:40-51, 1968.

63. Rimland D. Mechanisms of ethanol-induced defects of

alveolar macrophage function. Alcoholism: Clinical and Experimental Research 8(1):73-76, 1983.

64. Redei E; Clark WR; and McGivern RF. Alcohol exposure in utero results in diminished T-cell function and alterations in brain corticotropin-releasing factor and ACTH content. Alcoholism: Clinical and Experimental Research 13(3):439-443, 1989.

65. Johnson S.; Knight, R.; Marmier, D.J.; and Steele, R.W. Immunodeficiency in fetal alcohol syndrome. Pediatric Research 15(6):908-911, 1981.

66. Ewald SJ. T lymphocyte populations in fetal alcohol syndrome. Alcoholism: Clinical and Experimental Research 13(4):485-489, 1989.

67. The Impact of Sex Selection and Abortion in China, India and South Korea. Science Daily. 14 March 2011. http://www.sciencedaily.com/releases/2011/03/110314132244.htm <accessed 21 November 2011>.

68. Hvistendahl M. Unnatural Selection: Choosing Boys Over Girls, and the Consequences of a World Full of Men. New York: Public Affairs, 2011.

69. Douthat R. 160 Million and Counting. Opinion Pages. New York Times. 26 June 2011. http://www.nytimes.com/2011/06/27/opinion/27douthat.html?_r=1&partner=rssnyt&emc=rss <accessed 21 November 2011>.

70. Last JV. The War Against Girls. Wall Street Journal. 24 June 2011. http://online.wsj.com/article/SB10001424052702303

657404576361691165631366.html <accessed 21 November 2011>.

71. Mara Hvistendahl, http://marahvistendahl.com/index.php/book/ <accessed 21 November 2011>.

72. Josephine McDermott. A world full of men. Expat. http://my.telegraph.co.uk/expat/tag/mara-hvistendahl/ <accessed 21 November 2011>.

73. Pace DG. "HIV/AIDS and the Hispanic Community: Current Overview and Future Prospects." *The Year of Change.* National Association of Hispanic and Latino Studies. 2009. The authors express appreciation to the National Association of Hispanic and Latino Studies for allowing us to include material which Dr. Pace first presented at the 2009 conference, and then published in the proceedings of that conference.

74. Jerusalem J, The AIDS song, The Stigma song. http://www.hiv.co.il/aidssong/aidssong1.htm <accessed 7 December 2011>.

75. Barbara L. "Letters from Uganda: Children with AIDS feel despair." *Mormon Times*. 5 March 2011, p. 5.

76. Atwine B, Cantor-Graae E, Banjunirwe F. (March 2005), Psychological distress among AIDS orphans in rural Uganda. Social Science & Medicine 61 555-564.

77. USAID/SCOPE-OVC/FHI (2002), Results of the orphans and vulnerable children head of household baseline survey in four districts in Zambia. http://www.fhi.org/en/HIVAIDS/pub/Archive/OVCDocuments/ovczambia.htm <accessed 9 December 2011>.

78. Monasch R and Boerma JT (2004), Orphanhood and childcare

patterns in Sub-Saharan Africa: an analysis of national surveys from 40 countries. AIDS 18 (suppl. 2): S55-S65.

79. Salaam T., Congressional Research Service (2005), AIDS orphans and vulnerable children (OVC): problems, responses and issues for congress. http://www.law.umaryland.edu/marshall/crsreports/crsdocuments/RL32252112304.pdf <accessed 9 December 2011>.

80. IRIN News (October 2002), Botswana: AIDS orphans exploited. http://www.plusnews.org/ <accessed 9 December 2011>.

81. AVERT. AIDS Orphans. http://www.avert.org/aids-orphans.htm <accessed 9 December 2011>.

82. UNAIDS (2006) 'Report on the global AIDS epidemic. http://www.unaids.org/en/Dataanalysis/Epidemiology/ <accessed 9 December 2011>

83. United Nations Children's Fund (2003) Statement by UNICEF representative Bjorn Ljungqvist, HIV/AIDS orphans survey findings conference, April 8.

84. Centers for Disease Control and Prevention. HIV among Latinos. http://www.cdc.gov/hiv/latinos/index.htm <accessed 12 December 2011>.

85. Alanis MC and Lucidi RS (2004) "Neonatal Circumcision: A Review of the World's Oldest and Most Controversial Operation." Obstetrical & Gynecological Survey. 59(5):379-395.

86. Centers for Disease Control and Prevention. "Transmission categories and country of birth of Hispanics/Latinos with AIDS diagnosed in the United States during 2006," Centers

for Disease Control. http://www.cdc.gov/hiv/hispanics/ <accessed 20 January 2009>]

87. Auvert B, Taljaard D, Lagarde E, Sobngwi-Tambekou J, Sitta R, Puren A, *et al.* (2005) "Randomized, controlled intervention trial of male circumcision for reduction of HIV infection risk: The ANRS 1265 trial." PLoS Med. Nov;2(11):e298. Epub 2005 Oct 25. Erratum in: PLoS Med. 2006 May;3(5):e298.

88. Bailey RC, Moses S, Parker CB, Agot K, Maclean I, Krieger JN, *et al.* (2007) "Male circumcision for HIV prevention in young men in Kisumu, Kenya: a randomized controlled trial." Lancet 369:643-656.

89. Bongaarts J, Reining P, Way P, Conant F (1989) "The relationship between male circumcision and HIV infection in African populations." AIDS 3:373-377.

90. Ruiz, Sonia, Jennifer Kates, and Claire Oseran Pontius, Henry J. Kaiser Family Foundation, *Key Facts: Latinos and HIV/AIDS* (July 2003) http://www.kff.org/hivaids/6088-index.cfm. <accessed 9 January 2009>.

91. Centers for Disease Control (CDC). "HIV/AIDS, Hispanics/ Latinos." http://www.cdc.gov/hiv/hispanics/index.htm. <accessed 20 January 2009>.

92. Centers for Disease Control (CDC). "STDs in Minorities." http://www.cdc.gov/std/stats98/98pdf/Section9.pdf <accessed 20 January 2009>.

93. Donoval BA, Landay AL, Moses S, Agot K, Ndinya-Achola JO, Nyagaya EA, *et al.* (2006) "HIV-1 target cells in foreskins of African men with varying histories of sexually transmitted infections." Am J Clin Pathol 125: 386–91.

94. Drain P, Halperin D, Hughes J, Klausner J, Bailey R. (2006) "Male circumcision, religion, and infectious diseases: an ecologic analysis of 118 developing countries." BMC Infect Dis 6: 172.

95. Gollaher DL (2000) *Circumcision: a history of the world's most controversial surgery.* New York, NY: Basic Books, p53–72.

96. Kebaabetswe P, Lockman S, Mogwe S, Mandevu R, Thior I, Essex M. et. al. "Male circumcision: an acceptable strategy for HIV prevention in Botswana." Sexually Transmitted Infections 79 (2003): 214-219.

97. Morris, Brian J. "Male Circumcision Guide for Doctors, Parents, Adults & Teens." http://www.circinfo.net/rates_of_circumcision.html <accessed 5 February 2009>.

98. Williams BG, Lloyd-Smith JO, Gouws E, Hankins C, Getz WM, Hargrove J, et al. (2006) "The potential impact of male circumcision on HIV in sub-Saharan Africa." PLoS Med 3: e262.

99. "Subpopulation Estimates from the HIV Incidence Surveillance System --- United States, 2006" Morbidity and Mortality Weekly Report (MMWR) September 12, 2008 / 57(36);985-989 http://www.cdc.gov/mmwr/preview/mmwrhtml/mm5736a1.htm. <accessed 20 January 2009>.

100. World Health Organization (WHO), "Call for action to reduce the harmful use of alcohol," http://www.who.int/mediacentre/news/releases/2010/alcohol_20100521/en/index.html <accessed 12 December 2011>.

101. Nebehay S. "Alcohol kills more than AIDS, TB or

violence." Reuters. 13 February 2011. http://www.asiaone.com/Health/News/Story/A1Story20110213-263238.html <accessed 12 December 2011>

102. "A Snapshot of Annual High-Risk College Drinking Consequences." College Drinking—Changing the Culture" http://www.collegedrinkingprevention.gov/statssummaries/snapshot.aspx <accessed 12 December 2011>.

103. Hingson RW, Heeren T, Zakocs RC, Kopstein A, Wechsler H. Magnitude of alcohol-related mortality and morbidity among U.S. college students ages 18-24. *Journal of Studies on Alcohol* 63(2):136-144, 2002.

104. Hingson, R. et al. Magnitude of Alcohol-Related Mortality and Morbidity Among U.S. College Students Ages 18-24: Changes from 1998 to 2001. Annual Review of Public Health, vol. 26, 259-79; 2005

105. Hingson RW, Howland J. Comprehensive community interventions to promote health: Implications for college-age drinking problems. *Journal of Studies on Alcohol Supplement* 14:226-240, 2002.

106. Engs RC, Diebold BA, Hansen DJ. The drinking patterns and problems of a national sample of college students, 1994. *Journal of Alcohol and Drug Education* 41(3):13-33, 1996.

107. Presley CA, Meilman PW, Cashin JR. *Alcohol and Drugs on American College Campuses: Use, Consequences, and Perceptions of the Campus Environment, Vol. IV: 1992-1994.* Carbondale, IL: Core Institute, Southern Illinois University, 1996a.

108. Presley CA, Meilman PW, Cashin JR, Lyerla R.

*Alcohol and Drugs on American College Campuses: Use, Consequences, and Perceptions of the Campus Environment, Vol. III: 1991-1993.* Carbondale, IL: Core Institute, Southern Illinois University, 1996b.

109. Wechsler H, Lee JE, Kuo M, Seibring M, Nelson TF, Lee HP. Trends in college binge drinking during a period of increased prevention efforts: Findings from four Harvard School of Public Health study surveys, 1993-2001. *Journal of American College Health* 50(5):203-217, 2002.

110. Lönnroth K et al. (2008). Alcohol use as a risk factor for tuberculosis—a systematic review. BMC Public Health , 8:289.

111. Baliunas D et al. (2009). Alcohol consumption and risk of incident human immunodeficiency virus infection: a meta-analysis. International Journal of Public Health, 55:159–166 [Epub2009 Dec 1].

112. National Institute of Medical Statistics, National AIDS Control Organisation, Technical Report: India HIV Estimates-2006. http://www.unaids.org/en/dataanalysis/epidemiology/countryestimationreports/india_hiv_estimates_report_2006_en.pdf <accessed 4 November 2011>).

113. National Department of Health, South Africa. The National HIV and Syphilis Prevalence Survey: South Africa 2007. 2008. UNAIDS http://www.unaids.org/en/dataanalysis/epidemiology/countryestimationreports/20080904_southafrica_anc_2008_en.pdf<accessed 4 November 2011>).

114. Jayaprakash Institute of Social Change. Summary Report

of the Situation Analysis of Widows in Religious Places of Wet Bengal. New Delhi: Ministry of Women and Child Development, Government of India, 2009. http://wcd.nic.in/research/situanwidowswb.pdf <accessed 13 December 2011>.

115. Williams BG, Lloyd-Smith JO, Gouws E, Hankins C, Getz WM, et al. (2006) The Potential Impact of Male Circumcision on HIV in Sub-Saharan Africa. PLoS Med 3(7): e262. doi:10.1371/journal.pmed.0030262.

116. The Andhra Pradesh Devadasis (Prohibition of Dedicated) Act, 1988. Laws of India: A Project of PRS Legislative Research. http://www.lawsofindia.org/statelaw/2693/TheAndhraPradeshDevadasisProhibitionofDedicatedAct1988.html <accessed 13 December 2011>.

117. The Tamil Nadu Devadasis (Prevention of Dedication) Act, 1947. Laws of India: A Project of PRS Legislative Research. http://www.lawsofindia.org/pdf/tamil_nadu/1947/1947TN31.pdf <accessed 13 December 2011>.

118. Pace DG, Bagasra O. *Reflections on South Asia: Private Wants or Community Needs?* Saarbrücken: Germany. VDM Verlag Dr. Müller Aktiengesellschaft. 2010.

119. National AIDS Control Organisation. Antiretroviral Therapy Guidelines for HIV-Infected Adults and Adolescents Including Post-exposure Prophylaxis. Ministry of Health and Family Welfare, Government of India2007, with support from CDC,.Clinton Foundation, WHO. http://www.ilo.org/wcmsp5/groups/public/---ed_protect/---protrav/---ilo_aids/documents/legaldocument/wcms_117317.pdf <accessed 13 December 2011>.

120. Pace DG, Bagasra O. "NACO and the World Bank are correct in their crackdowns." *Nature Medicine* 14, 588 (2008). doi:10.1038/nm0608-588.

121. BBC News. "Alcohol 'aids HIV cell infection.'" http://news.bbc.co.uk/2/hi/health/4123193.stm <accessed 13 December 2011>.

122. CBS. "AIDS Out Of Control In India: Bill Gates Donating Millions To Help Stop Epidemic." 11 April 2004. http://www.cbsnews.com/stories/2004/04/08/60minutes/main610961.shtml <accessed 18 January 2008>.

123. HIV/AIDS and the Military. Plus News. http://www.irinnews.org/pdf/pn/Plusnews-Media-Fact-file-Military.pdf <accessed 13 December 2011>.

124. Rupiya. M (Editor), The Enemy Within: Southern African Militaries' Quarter-Century Battle with HIV and AIDS, Institute for Security Studies, 2006. http://www.iss.co.za/uploads/FULLPDFENEMYWITHIN.PDF <accessed 13 December 2011>.

125. Hofmeyr SA. An Interpretive Introduction to the Immune System. Forester Publication, 1999 and 2000.).

126. Frenkel N, Schirmer EC, Wyatt LS, Katsafanas G, Roffman E, Danovich RM, June CH. Isolation of a new herpesvirus from human CD4+ T cells. Proc. Natl. Acad. Sci. USA 1990; 87:748-752.

127. Bagasra O. HIV and Molecular Immunity: Prospect for AIDS Vaccine. Eaton Publishing, 1999; Natic, MA, USA.

128. Bagasra O. RNAi as an anti-viral therapy. Expert Opin Biol Ther.2005; 5(11):1463-1474.

129. Bagasra O, Pace DG. New Direction for HIV-1 Vaccine. Current Trends in Immunology, 2011; 2: 1-13.

130. Milush JM, Mir KD, Sundaravaradan V, Gordon SN, Engram J, Cano CA, Reeves JD, Anton E, O'Neill E, Butler E, Hancock K, Cole KS, Brenchley JM, Else JG, Silvestri G, Sodora DL. Lack of clinical AIDS in SIV-infected sooty mangabeys with significant CD4+ T cell loss is associated with double-negative T cells. J Clin Invest. 2011;121(3):1102-10. /

131. Bagasra O. A unified concept of HIV-1 latency. Expert Opin Biol Ther 2006; 6: 1135-1149.)

132. Chowdhury K, Bagasra O. Edible vaccine for malaria using transgenic tomatoes of varying sizes shapes and colors to carry different antigens. Medical Hypotheses 2007; 68:22-30.).

133. Cao Y, Qin L, Zhang L, Safrit J, Ho DD. Virologic and immunologic characterization of long-term survivors of human immunodeficiency virus type 1 infection. N. Engl. J. Med. 1995; 332:201-208.).

134. Bagasra O, Pace DG. Immunity and the Quest for an HIV Vaccine: A New Perspective. Bloomington, IN: AuthorHouse. 2011.

135. Medzhitov R, Littman D. HIV immunology needs a new direction. Nature. 2008; 455:591.

136. Stoiber H. Complement, Fc receptors and antibodies: a Trojan horse in HIV infection? Curr Opin HIV AIDS. 2009; 4(5):394-9.

137. Johnston MI, Fauci AS. An HIV vaccine-challenges and prospects. N Engl J Med. 2008; 359:888-90.

138. Watkins DI, Burton DR, Kallas EG, Moore JP, Koff WC. Nonhuman primate models and the failure of the Merck HIV-1 vaccine in humans. Nat Med. 2008; 14:617-21.

139. Steinbrook R. One step forward, two steps back--will there ever be an AIDS vaccine? New Engl J Med. 2007; 357:2653-5.

140. Pantaleo G. HIV-1 T-cell vaccines: evaluating the next step. Lancet Infect Dis. 2008; 8:82-3.

141. Munker R, Calin GA. MicroRNA profiling in cancer. Clin Sci (Lond). 2011 Aug;121(4):141-58.

142. Li B, Stefano-Cole K, Kuhrt DM, Gordon SN, Else JG, Mulenga J, Allen S, Sodora DL, G, Derdeyn CA. Nonpathogenic simian immunodeficiency virus infection of sooty mangabeys is not associated with high levels of autologous neutralizing antibodies. J Virol. 2010;84(12):6248-53.

143. Apetrei C, Robertson DL, and Marx PA. The history of SIVS and AIDS: epidemiology, phylogeny and biology of isolates from naturally SIV infected non-human primates (NHP) in Africa. Front. Biosci. 2004; 9:225–254.

144. Vinton C, Klatt NR, Harris LD, Briant JA, Sanders-Beer BE, Herbert R, Woodward R, Silvestri G, Pandrea I, Apetrei C, Hirsch VM, Brenchley JM. CD4-Like Immunological Function by CD4- T Cells in Multiple Natural Hosts of Simian Immunodeficiency Virus. J Virol. 2011 Sep;85(17):8702-8.

145. Apetrei C, Gormus B, Pandrea I, Metzger M, ten Haaft

P, Martin LN, Bohm R, Alvarez X, Koopman G, Murphey-Corb M, Veazey RS, Lackner AA, Baskin G, Heeney J, Marx PA. Direct inoculation of simian immunodeficiency virus from sooty mangabeys in black mangabeys (Lophocebus aterrimus): first evidence of AIDS in a heterologous African species and different pathologic outcomes of experimental infection. J. Virol. 2004; 78:11506–11518.

146. Levine AL. Why do we not yet have a human immunodeficiency virus vaccine? J Virol. 2008; 82:11998-12000.

147. Anson DS. The use of retroviral vectors for gene therapy-what are the risks? A review of retroviral pathogenesis and its relevance to retroviral vector-mediated gene delivery. Genet Vaccines Ther. 2004; 2(1):9.

148. Anderson C. AIDS research. New findings cast doubt on UK vaccine trials. Nature. 1991; 353(6342):28.

149. Putkonen P, Nilsson C, Hild K, Benthin R, Cranage M, Aubertin AM, Biberfeld G. Whole inactivated SIV vaccine grown on human cells fails to protect against homologous SIV grown on simian cells. J Med Primatol. 1993; 22(2-3):100-3.

150. Hartung S, Norley SG, Ennen J, Cichutek K, Plesker R, Kurth R. Vaccine protection against SIVmac infection by high- but not low-dose whole inactivated virus immunogen. J Acquir Immune Defic Syndr. 1992; 5(5):461-8.

151. Mansfield K, Lang SM, Gauduin MC, Sanford HB, Lifson JD, Johnson RP, Desrosiers RC. Vaccine protection by live, attenuated simian immunodeficiency virus in the absence of high-titer antibody responses and high-frequency cellular

immune responses measurable in the periphery. J Virol. 2008; 82(8):4135-48.

152. Almond N, Jenkins A, Jones S, Arnold C, Silvera P, Kent K, Mills KH, Stott EJ. The appearance of escape variants in vivo does not account for the failure of recombinant envelope vaccines to protect against simian immunodeficiency virus. J Gen Virol. 1999; 80 ( Pt 9):2375-82.

153. Hulate SL, Cale EM, Korber BT, Letvin NL. Vaccine-induced CD8+ T lymphocytes of rhesus monkeys recognize variant forms of an HIV epitope but do not mediate optimal functional activity. J Immunol. 2011 May 15;186(10):5663-74.

154. Furesz J, Levenbook I. New assays for the quality control of live oral p

155. Lauring AS, Jones JO, Andino R. Rationalizing the development of live attenuated virus vaccines. Nat Biotechnol. 2010; 28(6):573-9.).

156. Cardo DM, Culver DH, Ciesielski CA, Srivastava PU, Marcus R, Abiteboul D, Heptonstall J, Ippolito G, Lot F, McKibben PS, Bell DM. A case-control study of HIV sero-conversion in health care workers after percutaneous exposure. Centers for Disease Control and Prevention Needlestick Surveillance Group. N Engl J Med. 1997; 337:1485-90.

157. Lusso P, Secchiero P, Crowley RW, Garzino-Demo A, Berneman ZN, Gallo RC. CD4 is a critical component of the receptor for human herpesvirus 7: Interference with human immunodeficiency virus. Proc. Natl. Acad. Sci. USA 1994; 91:3872-3876.

158. Kanak MA, Alseiari MA, Addanki KC, Aggarwal M, Noorali S, Kalsum A, Mahalingam K, Panasik N, Pace DG, Bagasra O. Triplex Forming microRNAs Form Stable Complexes with HIV-1 provirus and Inhibit Its Replication. Appl Immunohistochem Mol Morphol. 2010;18(6):532-4.

159. Szathmáry E. The origin of the genetic code: amino acids as cofactors in an RNA world. Trends Genet. 1999; 15:223-9. /

160. Bagasra O, Pace DG. Back to the Soil: Retroviruses and Transposons. In Biocommunication of soil-bacteria and viruses. Guenther Witzany, Ed. Chapter 6. Springer Press 2010;pages 161-188). /

161. Shabalina SA, Koonin EV. Origins and evolution of eukaryotic RNA interference. Trends Ecol Evol. 2008; 23(10):578-87.).

162. Schluter SF, Marchalonis JJ. Cloning of shark RAG2 and characterization of the RAG1/RAG2 gene locus. FASEB J. 2003; 17:470-2. /

163. Helm T. Basic immunology: a primer. Minn Med. 2004; 87(5):40-4.).

164. Pandrea I, Sodora DL, Silvestri G, Apetrei C. Into the wild: simian immunodeficiency virus (SIV) infection in natural hosts. Trends Immunol. 2008; 29:419-28. /

165. Bagasra O, Stir AE, Pirisi-Creek L, Creek KE, Bagasra AU, Glenn N, Lee JS. Role of Micro-RNAs in Regulation of Lentiviral Latency and Persistence. Applied Immunohistochemistry & Molecular Morphology. 2006; 14:276-290.

166. Goh WC, Markee J, Akridge RE, Meldorf M, Musey L, Karchmer T, Krone M, Collier A, Corey L, Emerman M, McElrath MJ. Protection against human immunodeficiency virus type 1 infection in persons with repeated exposure: evidence for T-cell immunity in the absence of inherited CCR5 coreceptor defects. J. Infect. Dis. 1999; 179:548-557.

167. Booth W. AIDS and insects. Science. 1987; 237(4813):355-6.

168. Duesberg PH. Inventing the AIDS virus. Regency Publishing Co., Inc. Washington D.C. 1997; pp 169-216.

169. Dowling T. "Uqedisizwe–The Finisher of the Nation." HIV/AIDS and African Languages. 2006. http://www.africanvoices.co.za/research/aidsresearch.htm <accessed 15 December 2011>.

170. Johnson B. "Top 10 Notable People Who Died From AIDS." http://listverse.com/2011/12/01/top-10-notable-people-who-died-from-aids/ <accessed 14 December 2011>.]

171. Gunn J, Jenkins J. Freddie Mercury: biography. http://www.freddie.ru/e/bio/ <accessed 14 December 2011>.

172. Morrison A. Arthur Ashe. Biography. CNNSI. http://sportsillustrated.cnn.com/tennis/features/1997/arthurashe/biography1.html <accessed 14 December 2011>.

173. Mandela's eldest son dies of Aids. BBC News. 6 January 2005. http://news.bbc.co.uk/2/hi/africa/4151159.stm <accessed 14 December 2011>.

174. "Mandela's son dies of AIDS at 54." Reuters. http://www.

utexas.edu/conferences/africa/ads/233.html <accessed 14 December 2011>.

175. Aids kills Zulu leader's daughter. BBC News. 7 August 2004. http://news.bbc.co.uk/2/hi/africa/3545450.stm <accessed 15 December 2011>.

176. Isaac Asimov. http://www.asimovonline.com/asimov_home_page.html <accessed 14 December 2011>.

177. Pace DG. "From Nimby Problems to Nioby Complexes: Wicked Webs as International Policy Dilemmas." In Greg A. Phelps, *Cross Currents: Renewable Energy Use, 1997-1998.* Clayton, Missouri: The Alan Company, 1997, pp. 5-39.

178. Skar H, Hedskog C, Albert J. HIV-1 evolution in relation to molecular epidemiology and antiretroviral resistance. Ann. N.Y. Acad. Sci. 1230 (2011) 108-118.

179. Gray RH, Kigozi G, Serwadda D, Makumbi F, Watya S, Nalugoda F, Kiwanuka N, Moulton LH, Chaudhary MA, Chen MZ, Sewankambo NK, Wabwire-Mangen F, Bacon MC, Williams CF, Opendi P, Reynolds SJ, Laeyendecker O, Quinn TC, Wawer MJ. 2007. Male circumcision for HIV prevention in men in Rakai, Uganda: a randomised trial. Lancet369: 657–666.

180. Wawer MJ, Makumbi F, Kigozi G, Serwadda D, Watya S, Nalugoda F, Buwembo D, Ssempijja V, Kiwanuka N, Moulton LH, Sewankambo NK, Reynolds SJ, Quinn TC, Opendi P, Iga B, Ridzon R, Laeyendecker O, Gray RH. 2009. Circumcision in HIV-infected men and its effect on HIV transmission to female partners in Rakai, Uganda: a randomised controlled trial. Lancet374: 229–237.

181. Wang L, Wang N. HIV/AIDS epidemic and the development of comprehensive surveillance system in China with challenges. Chin Med J (Engl). 2010 Dec;123(23):3495-500.

182. Gupta S, Boojh R. (2006) Report: biomedical waste management practices at Balrampur Hospital, Lucknow, India. Waste Manag Res. 24:584-91.

183. Singh S, Dwivedi SN, Sood R, Wali JP. (2000) Hepatitis B, C and human immunodeficiency virus infections in multiply-injected kala-azar patients in Delhi. Scand J Infect Dis. 32:3-6.

184. Sood A, Midha V, Awasthi G. (2002) Hepatitis C – knowledge and practices among the family physicians. Trop Gastroenterol. 23:198-201.

185. Wig N. (2003) HIV: awareness of management of occupational exposure in health care workers. Indian J Med Sci. 57:192-8.

186. 186. Murhekar MV, Rao RC, Ghosal SR, Sehgal SC. (2005) Assessment of injection-related practices in a tribal community of Andaman and Nicobar Islands, India. Public Health. 119:655-8.

187. Sumner, William Graham, War and Other Essays, pp. 176-177. Cited by Jonathan Marshall, William Graham Sumner: Critic of Progressive Liberalism. The Journal of Libertarian Studies 265. (http://www.mises.org/journals/jls/3_3/3_3_2.pdf <accessed 13 March 2009>).

188. Cannon, Lou. (2000) President Reagan: The Role of a Lifetime. New York: Public Affairs.

189. Human Rights Watch. Future Forsaken (2004) Abuses Against Children Affected by HIV/AIDS in India. New York.

190. Rajkumar, TS. (2004) Community towards the Needle syringe exchange program. International Conference on AIDS, Bangkok, Thailand. Int Conf AIDS. 2004 Jul 11-16; 15: abstract no. ThPeD7742).

191. Stine GJ. (2009) AIDS update 2008. McGraw Hill, 235-6.

192. WB upset at corruption taking toll on its health projects. The Economic Times, 12 January 2008. http://economictimes.indiatimes.com/News/Economy/Finance/WB_upset_at_corruption_taking_ toll_on_its_health _projects/articleshow/2693961.cms <accessed 13 March 2009>.

193. Priddy F, Tesfaye F, Mengistu Y, Rothenberg R, Fitzmaurice D, Mariam DH, de Rio C, Oli K, Worku A. (2005) Potential for medical transmission of HIV in Ethiopia. AIDS. 19:348-50; AIDS. 2006 Jan 2;20(1):133; author reply 133-5.

194. Hutin YJ, Hauri AM, Armstrong GL. (2003) Use of injections in healthcare settings worldwide, 2000: literature review and regional estimates. BMJ. 327(7423):1075.

195. Biswas D, Borkakoty BJ, Mahanta J, Jampa L, Deouri LC. (2007) Hyperendemic foci of hepatitis B infection in Arunachal Pradesh, India. J Assoc Physicians India. 55:701-4.

196. Kermode M, Longleng V, Singh BC, Hocking J, Langkham B, Crofts N. (2007) My first time: initiation into injecting

drug use in Manipur and Nagaland, north-east India. Harm Reduct J. 5:4:19.

197. Panda S, Sharma M. (2006) Needle syringe acquisition and HIV prevention among injecting drug users: a treatise on the "good" and "not so good" public health practices in South Asia. Subst Use Misuse. 41:953-77.

198. Bagasra, O, Pace DG. (2008) NACO and the World Bank are correct in their crackdowns. Nature Medicine. 14.6:588.

199. National AIDS Control Organisation (NACO), Ministry of Health and Family Welfare, Government of India, New Delhi. Progress Report on the Declaration of Commitment on HIV/AIDS, United Nations General Assembly Special Session (UNGASS) on HIV/AIDS, 2005. http://data.unaids.org/pub/Report/2006/2006_country_progress_report_india_en.pdf <accessed 13 March 2009>.

200. Padma, T.V. (2008) India continues crackdown, dismissing hundreds of AIDS groups. Nature Medicine. 14:227.

201. Rodrigues C, Ghag S, Bavi P, Shenai S, Dastur F. (2005) Needlestick injuries in a tertiary care centre in Mumbai, India. J Hosp Infect. 60:368-73.

202. Zambrano, María. La confesión: género literario. Madrid: Ediciones Siruela, (1943) 1995.

203. Pace, Donald Gene. 2009. Unfettering Confession: Ritualized Performance in Spanish Narrative and Drama. Lanham, MD. University Press of America.

204. Samet JH, Pace CA, Cheng DM, Coleman S, Bridden C, Pardesi M, Saggurti N, Raj A. Alcohol use and sex risk behaviors among HIV-infected female sex workers (FSWs)

and HIV-infected male clients of FSWs in India. AIDS Behav. 2010 Aug;14 Suppl 1:S74-83.

205. Kumar R, Perez-Casanova AE, Tirado G, Noel RJ, Torres C, Rodriguez I, Martinez M, Staprans S, Kraiselburd E, Yamamura Y, Higley JD, Kumar A. Increased viral replication in simian immunodeficiency virus/simian-HIV-infected macaques with self-administering model of chronic alcohol consumption. J Acquir Immune Defic Syndr. 2005 Aug 1;39(4):386-90.

206. Bagby GJ, Zhang P, Purcell JE, Didier PJ, Nelson S. Chronic binge ethanol consumption accelerates progression of simian immunodeficiency virus disease. Alcohol Clin Exp Res. 2006 Oct;30(10):1781-90.

207. Poonia B, Nelson S, Bagby GJ, Zhang P, Quniton L, Veazey RS. Chronic alcohol consumption results in higher simian immunodeficiency virus replication in mucosally inoculated rhesus macaques. AIDS Res Hum Retroviruses. 2006 Jun;22(6):589-94.

208. Bagasra O, Kajdacsy-Balla A, Lischner HW, Pomerantz RJ. Alcohol intake increases human immunodeficiency virus type 1 replication in human peripheral blood mononuclear cells. J Infect Dis. 1993 Apr;167(4):789-97.

# About the Authors

Dr. Donald Gene Pace (two PhDs; a highly published writer that examines public health policy, history, and literature) and Dr. Omar Bagasra (MD, PhD; an eminent molecular biologist, immunologist, and retrovirologist) are professors at Claflin University (Orangeburg, South Carolina). They are the authors of many articles or book chapters, and have previously coauthored other books, including Reassessing HIV Vaccine Design and Approaches: Towards a Paradigm Shift (2013), Immunology and the Quest for an HIV Vaccine: A New Perspective (2011) and Public Policy in South Asia: Private Wants or Community Needs? (2010).